MOLECULAR MARINE MICROBIOLOGY

JMMB Symposium Series Volume 1

The reviews published in this book were originally presented as a written symposium in the Journal of Molecular Microbiology and Biotechnology Vol. 1 No. 1 (August 1999)

Edited by

Douglas H. Bartlett
UCSD, La Jolla, California

 horizon scientific press

Copyright © 2000
Horizon Scientific Press
32 Hewitts Lane
Wymondham
Norfolk NR18 0JA
England

www.horizonpress.com

British Library Cataloguing-in-Publication Data

A catalogue record for this book is available from the British Library

ISBN: 1-898486-20-4 ✓

*Printed and bound in Great Britain
by Biddles Ltd, Guildford and King's Lynn*

Contents

1

Introduction

Douglas H. Bartlett

Center for Marine Biotechnology and Biomedicine
Scripps Institution of Oceanography
University of California, San Diego
La Jolla, CA 92093-0202, USA

The following thirteen chapters introduce some of the exciting milestones being covered in molecular marine microbiology today, and in many cases, their biomedical or biotechnological relevance. The issues covered include quorum and anti-quorum sensing, selected symbioses, DNA shuffling in the environment, the evolution and development of novel motility mechanisms, hydrocarbon degradation, the interactions of microbes with metals, and adaptations to extremes of pressure and temperature. Although most of these topics are not exclusive to marine realms, they all have as a common thread a profound relevance to the ocean environment.

One of the great contributions of marine microbiology to all of microbiology has been the discovery of the importance of acylated homoserine lactones in cell density or "quorum" sensing by bacteria. The paradigm for this mode of regulation, the control of bioluminescence in *Vibrio fischeri*, is described by Paul Dunlap. An entirely new light on *V. fischeri* is provided by Ned Ruby, who describes how this microorganism, along with a squid host with which it has established a mutualistic association, is providing new clues to bacterial colonization of animal epithelial tissue. A fascinating twist on quorum sensing is presented by Rice *et al.* who describe the synthesis of an inhibitor of bacterial quorum sensing systems by a species of red algae. The inhibitor appears to represent a chemical strategy used by the algae to limit bacterial colonization of its surfaces. Additional aspects of bioactive metabolites involved in the symbiotic interactions of marine microorganisms is described by Haygood *et al.* who describe potential biomedical applications of microbes obtained from marine invertebrates.

Of course microbes take in much more than cell-cell signaling molecules. Many prokaryotes can also take in DNA, and thus re-tailor their genetic identies. John Paul describes gene transfer among indiginous microbes in the environment and the significance of transformation, conjugation and transduction to phylogenetics. Additional microbial processing reactions which are of biotechnological relevance concern the applications of marine prokaryotes to environmental clean up and nano-scale biofabrication. Harayama *et al.* describe the fate of various petroleum components in seawater, and the significance of *Alcanivorax* in marine bioremediation. Francis and Tebo explore the details of manganese oxidation, progressing from genes to enzyme kinetics and applications to toxic metal removal. Finally, Dirk Schüler describes the bacterial manufacturing of magnetic iron containing organelles (magnetosomes) within certain benthic bacteria and their biotechnological promise in products ranging from magnetic tapes to magnetic resonance imaging devices.

Two chapters deal with mysterious motility mechanisms operating in marine microbes. Linda McCarter introduces surface-regulated phenomena, including swarming motility via lateral flagella, in the human pathogen and marine resident, *V. parahaemolyticus*. Bianca Brahamsha details the only known case of liquid motility in the absence of flagella, a phenotype displayed by many marine *Synechococcus* species.

The book ends with descriptions of the adaptations of certain deep-sea prokaryotes to extremes of pressure and temperature. Kato and Qureshi present a taxomonic overview of deep-sea bacteria and genetic and biophysical analyses of respiratory chain adaptations to elevated pressure. My article describes genetic evidence for the role of additional membrane components in pressure-sensing and adaptation, the discovery of a DNA recombination enzyme which is also required for life at low volume change, and the potential of deep-sea bacteria as a source of genes for the biosynthesis of polyunsaturated fatty acids. Finally, Robb and Clark present a discussion of the biochemical bases of thermostability in proteins from hyperthermophiles, most of which are members of the domain Archaea.

Although this admittedly ecclectic collection of articles is hardly representative of all of molecular marine microbiology (for example, no articles dealing with viruses or protists are included), the book provides a valuable glimpse into many of the exciting developments taking place in this broad field. Readers will find the book useful both as a reference source and as an introduction to topics yet to be explored.

2

Quorum Regulation of Luminescence in *Vibrio fischeri*

Paul V. Dunlap

Center of Marine Biotechnology, University of Maryland
Biotechnology Institute, Columbus Center, Suite 236,
701 East Pratt Street, Baltimore, Maryland 21202, USA

Abstract

Luminescence in *Vibrio fischeri* is controlled by a population density-responsive regulatory mechanism called quorum sensing. Elements of the mechanism include: LuxI, an acyl-homoserine lactone (acyl-HSL) synthase that directs synthesis of the diffusible signal molecule, 3-oxo-hexanoyl-HSL (*V. fischeri* auto-inducer-1, VAI-1); LuxR, a transcriptional activator protein necessary for response to VAI-1; GroEL, which is necessary for production of active LuxR; and AinS, an acyl-HSL synthase that catalyzes the synthesis of octanoyl-HSL (VAI-2). The population density-dependent accumulation of VAI-1 triggers induction of *lux* operon (*luxICDABEG*; genes for luminescence enzymes and for LuxI) transcription and luminescence by binding to LuxR, forming a complex that facilitates the association of RNA polymerase with the *lux* operon promoter. VAI-2, which apparently interfers with VAI-1 binding to LuxR, operates to limit premature *lux* operon induction. Hierarchical control is imposed on the system by 3':5'-cyclic AMP (cAMP) and cAMP receptor protein (CRP), which are necessary for activated expression of *luxR*. Several non-*lux* genes in *V. fischeri* are controlled by LuxR and VAI-1. Quorum regulation in *V. fischeri* serves as a model for LuxI/LuxR-type quorum sensing systems in other Gram-negative bacteria.

Introduction

Over the past few years, awareness of a new paradigm in bacterial gene regulation, called quorum sensing, has developed. Classically defined, quorum sensing involves the production, via proteins of the LuxI family, of an acyl-homoserine lactone (acyl-HSL) signal molecule and the response to that signal via transcriptional regulatory proteins of the LuxR family (Fuqua *et al.*, 1996; Dunlap, 1997; Hastings and Greenberg, 1999). There is also growing recognition of quorum sensing, via other types of signals and other types of regulatory proteins, in Gram-positive bacteria and potentially also in eukaryotic organisms (Kleerebezem *et al.*, 1998; Gomer, 1999). Originating primarily in studies of luminescence autoinduction in marine bacteria (Greenberg, 1997), LuxI/LuxR type quorum sensing systems have now been identified in over 25 species of Gram-negative bacteria, including several plant and animal pathogens. The activities regulated by quorum sensing include, for example, luminescence, conjugative plasmid transfer, and the production of antibiotics and extracellular enzymes. The growing number of species identified as using quorum sensing and the chemical and genetic homologies of the quorum sensing systems indicate that quorum regulation is an evolutionarily conserved mechanism for regulating gene expression that is widespread among bacteria. One of the first and presently the most thoroughly studied of the LuxI/LuxR type quorum sensing systems is that of *Vibrio fischeri*. Studies of luminescence in *V. fischeri* have been pursued in many laboratories over the past thirty years, and by defining the major themes and working out the details of quorum regulation in this bacterium, those studies have generated a physiological, chemical and molecular-genetic foundation for the identification and analysis of quorum sensing systems in other bacteria. New information continues to emerge on the *V. fischeri* system, which serves as a guide and a reference point for studies of quorum regulation in other species. This chapter provides an overview and current perspective on the quorum sensing system of *V. fischeri*.

Luminescence Induction

Light production by *V. fischeri* cells changes in a striking manner during batch culture growth. Initially high if strongly luminous cells are used as the inoculum, luminescence drops many fold during the first several cell divisions and then begins a sharp rise as the culture reaches late exponential phase, attaining a peak in early stationary phase. Synthesis of luciferase, the light-producing enzyme, generally follows the pattern of light production, with luciferase activity remaining at a fairly constant level in the culture for the first several cell divisions and then inducing rapidly as the culture enters late exponential phase. Experimental analysis of this dramatic pattern of light production in *V. fischeri* and *Vibrio harveyi* led to demonstrations that the cells produce and release into the medium a diffusible luminescence inducing factor, autoinducer, which accumulates during growth and triggers induction of luciferase when it attains a threshold concen-tration (Nealson *et al.*, 1970; Eberhard, 1972; Nealson, 1977; Nealson and Hastings, 1979; Rosson and Nealson, 1981). Originally called autoinduction to reflect the self-produced nature of the inducing factor, this form of regula-tion is now also referred to as quorum sensing (Fuqua *et al.*, 1994) to reflect its relationship to population density.

An Ecological Rationale for Quorum Regulation

Quorum sensing apparently functions to enable *V. fischeri* cells to discriminate between different habitats and adaptively produce or not produce light. In seawater, where the population density of *V. fischeri* cells is low ($<10^2$ cells/ml; Ruby and Nealson, 1978; Ruby *et al.*, 1980; Baumann and Baumann, 1981; Lee and Ruby, 1992), diffusion would not allow the inducer to accumulate to the threshold concentration necessary for induction (Rosson and Nealson, 1981). Presumably, luminescence would not be beneficial under these conditions, if one assumes that the function of luminescence is to be seen by higher organisms or is in some other way related to association with higher organisms (see Dunlap, 1991, for a discussion); the light produced by fewer than 100 cells dispersed throughout 1 ml of seawater, even if those cells are fully induced, presumably would be too weak to be seen by animals. In light organ symbiosis with fish and squid, where the density of *V. fischeri* cells is high (ca. 10^9 to 10^{10} cells/ml; Dunlap and Greenberg, 1991; Ruby, 1996), autoinducer concentrations exceed those necessary to induce luminescence (Ruby and Nealson, 1976; Ruby and Asato, 1993; Boettcher and Ruby, 1995), which is produced at a sufficiently high level to be used by the animal in visually related behaviors (*e.g.*, Singley, 1983; McFall-Ngai, 1990; Dunlap and Greenberg, 1991). Because luminescence requires energy, both for the luminescence reaction and for synthesis of luciferase (see Dunlap and Greenberg, 1991, for a discussion), it is logical to assume that luminescence would be induced only under those conditions when it enhances the survival or growth of *V. fischeri*. This notion implies that light production has a benefit for the bacteria, one that offsets its cost, and that a high level of light production may be in some way physiologically adaptive in the symbiotic state (Dunlap, 1985).

From the perspective of the host squids and fish that harbor *V. fischeri* or other luminous bacteria and use the bacterial light in luminous displays associated with predation, antipredation and intraspecific signaling (Hastings and Nealson, 1981; McFall-Ngai and Dunlap, 1983; Dunlap and Greenberg, 1991; McFall-Ngai, 1990), the benefit of bacterial light production for the host is clear. However, whether luminous bacteria provide other metabolic or nutritional benefits to their hosts, as seen in mutualisms between other kinds of symbiotic bacteria and their host animals has been a long-standing question. For example, it was not previously known whether luminous bacteria forming any of the various light organ mutualisms with squids and fishes favorably influenced or were required for the growth, survival and overall development of the host animal. Also unknown was the extent to which the light organ as a whole and its individual tissues would develop in the absence of the symbiotic bacteria. Recent success in culturing one of the hosts of *V. fischeri*, the sepiolid squid *Euprymna scolopes*, through its life cycle under laboratory conditions, has permitted a resolution of these issues (Claes and Dunlap, 1999). Animals colonized by *V. fischeri* and animals kept in an aposymbiotic state from the time of hatching were found to survive and grow equally well to adulthood, to develop similarly, to behave similarly at each developmental stage, and to reach sexual maturity at a similar age. Furthermore, overall development of the light organ and its accessory tissues (lens, reflectors and ink sac) in colonized and aposymbiotic animals exhibited no major morphometric or histological differences (Claes and Dunlap, 1999). In revealing that *V. fischeri* is not required for normal growth and development of the animal or for development of the light organ, the results of this study demonstrate that the metabolic contribution of the symbiotic bacterium to the host animal in this association apparently is limited to light production. These considerations place additional emphasis on the importance of quorum sensing regulation of luminescence for the symbiosis of *V. fischeri* with its squid and fish hosts.

Two Major Advances

Studies of luminescence regulation in *V. fischeri* during the early 1980's resulted in two major advances, the chemical identification of the inducing factor (Eberhard *et al.*, 1981) and the cloning of the luminescence genes (Engebrecht *et al.*, 1983). These advances opened up the quorum sensing mechanism in *V. fischeri* to detailed molecular and genetic analysis, and as a consequence they established the foundation for the discovery of quorum sensing systems in other bacteria. Eberhard *et al.* (1981) extracted and purified the autoinducing compound from medium conditioned by the growth of *V. fischeri*, identified it as 3-oxohexanoyl-L-homoserine lactone (*V. fischeri* autoinducer-1, VAI-1) and demonstrated that the chemically synthesized compound has full biological activity. Initially carefully conserved and doled out sparingly, VAI-1 recently became commercially available at low cost, reflecting its importance as a tool in quorum sensing research.

Shortly after VAI-1 was chemically identified, Engebrecht *et al.* (1983) isolated a fragment of the *V. fischeri* MJ-1 chromosome that conferred on *Escherichia coli* the ability to produce light and to regulate light production in a population density-responsive manner. The fragment was found to contain genes for the luminescence proteins and genes necessary for the regulated expression of those proteins, including a gene necessary and sufficient for synthesis of the *V. fischeri* autoinducer by *E. coli*. The *V. fischeri lux* genes are organized in two divergently transcribed units, *luxR* and *luxICDABEG* (the *lux* operon) (Engebrecht and Silverman, 1984). The first gene of the *lux* operon, *luxI*, and the single gene of the divergent transcriptional unit, *luxR*, are regulatory. *luxI* was defined as required for cells to synthesize VAI-1, whereas *luxR* specifies a protein necessary for cells to activate *lux* operon transcription in response to VAI-1 (Engebrecht *et al.*, 1983; Engebrecht and Silverman, 1984). The *luxA* and *luxB* genes specify the α and β subunits of luciferase, and *luxC*, *luxD*, and *luxE* encode the polypeptides of the *lux*-specific fatty acid reductase complex (reductase, acyl transferase, and acyl protein synthetase), which are necessary for the synthesis and recycling of the aldehyde substrate for luciferase (Boylan *et al.*, 1985; 1989). The last *lux* operon gene, *luxG*, which is followed by a strong transcriptional terminator (Swartzman *et al.*, 1990), encodes a probable flavin reductase of the Fre/LuxG family of NAD(P)H-flavin oxidoreductases (Zenno and Saigo, 1994). The LuxG protein is not required for high levels of light production in *V. fischeri* (A. Kuo and P.V. Dunlap, unpublished data), so another flavin reductase activity in this species (Zenno *et al.*, 1994) apparently can compensate for the loss of *luxG*.

The *V. fischeri lux* Regulatory Region

The *luxR* gene and the *lux* operon are separated by a 219 bp regulatory region that contains the promoters for these two transcriptional units. The *luxR* promoter consists of -10 and -35 regions similar to consensus sequences for *E. coli* "housekeeping" promoters and a consensus binding site for 3':5'-cyclic AMP (cAMP) receptor protein (CRP) at position -59 from the *luxR* transcriptional start. The *lux* operon promoter has a -10 region but has no -35 region identifiable by comparison with *E. coli* consensus sequences. At this position, centered 40 bp upstream of the *lux* operon transcriptional start, is a 20-bp region of dyad symmetry, termed the *lux* box. This region is required for activation of *lux* operon transcription and has been implicated as the binding site for LuxR (Baldwin *et al.*, 1989; Devine *et al.*, 1988; 1989; Engebrecht and Silverman, 1987; Shadel *et al.*, 1990; Stevens *et al.*, 1994; Stevens and Greenberg, 1997; Egland and Greenberg, 1999). Identical or similar sequences have been identified in the *lux* operon promoters of various *V. fischeri* strains as well as in the promoters controlling expression of *lasB* and *rhlI* in *Pseudomonas aeruginosa*, *traA* and *traI* of the octopine-type Ti plasmid, and *traI* of the nopaline-type plasmid, respectively, of *Agrobacterium tumefaciens*, *solI* of *Ralstonia solanacearum*, and *cepI* of *Burkholderia cepacia* (Engebrecht and Silverman, 1987; Devine *et al.*, 1988; 1989; Shadel *et al.*, 1990; Fuqua *et al.*, 1994; Fuqua and Winans, 1994; Gray *et al.*, 1994; Hwang *et al.*, 1994; Latifi *et al.*, 1995; Flavier *et al.*, 1997;

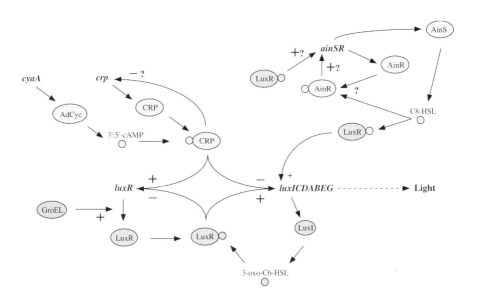

Figure 1. Regulatory Elements and Circuitry Controlling Luminescence in *V. fischeri*
Along with the quorum sensing system, consisting of LuxR, LuxI, and AinS, hierarchical control by cAMP/ CRP is also depicted, as is the influence of GroEL on LuxR. Abbreviations: AdCyc, adenylate cyclase; 3-oxo-C6-HSL, 3-oxo-hexanoyl-homoserine lactone; C8-HSL, octanoyl-homoserine lactone. Transcriptional activation and repression are indicated by + and -, respectively.

Lewenza *et al.*, 1999). The *lux* box represents a conserved regulatory sequence; its presence upstream of a bacterial gene is interpreted as consistent with autoinducer-mediated control of that gene (Gray *et al.*, 1994). The presence in *V. fischeri* of a *lux* box preceeding each of several recently isolated non-*lux* genes that are positively regulated by LuxR and VAI-1 (Callahan and Dunlap, manuscript in preparation) supports that interpretation.

A Molecular Model for Quorum Regulation of *lux* Operon Expression

The organization and function of the *lux* genes described above leads to a simple model for quorum regulation of the *lux* operon. A central aspect of the model is the membrane-permeant nature of VAI-1; the concentration of VAI-1 within cells rapidly equilibrates with that outside cells (Kaplan and Greenberg, 1985). According to the model, when *V. fischeri* cells fully induced for luminescence are transferred at a low population density to fresh medium, or in nature are released, for example, from the light organ of a fish or squid into seawater, VAI-1 rapidly diffuses away from the cells. The cellular concentration of VAI-1 then drops below the threshold necessary for activation of *lux* operon transcription, and synthesis of the *lux* operon proteins decreases sharply. As the population of *V. fischeri* cells grows in the medium, or increases following colonization of a squid or fish light organ (*e.g.*, Ruby, 1996), VAI-1 gradually accumulates within the medium and within cells. Presumably, de novo synthesis of VAI-1 results from the activity of extant LuxI protein and from a small amount of new LuxI produced from basal, unactivated *lux* operon transcription, with the test tube or animal light organ tubule providing a diffusion barrier. When VAI-1 attains the threshold level for interaction with LuxR, the LuxR/VAI-1 complex then interacts with RNA polymerase, facilitating its association with the *lux* operon promoter and transcription of the *lux* operon (Engebrecht and Silverman, 1984; Stevens *et al.*, 1994; Stevens and Greenberg, 1997) (Figure 1). An autocatalytic increase in synthesis of VAI-1 ensues, since *luxI* is part of the *lux* operon (Engebrecht *et al.*, 1983; Friedrich and Greenberg, 1983; Eberhard *et al.*, 1991), leading to a rapid increase in luciferase synthesis and luminescence.

With respect to synthesis of LuxR, transcription from the *luxR* promoter is activated by cAMP and CRP (described below). High levels of LuxR protein and VAI-1 apparently could place a cap on *lux* operon expression. VAI-1 and LuxR repress *luxR* expression post-transcriptionally and transcriptionally (Engebrecht and Silverman, 1986; Dunlap and Greenberg, 1988; Dunlap and Ray, 1989; Shadel and Baldwin, 1991; 1992a). Conversely, *luxR* expression can be activated by low levels of VAI-1 and LuxR in both a cAMP/CRP-dependent and cAMP/CRP-independent manner (Shadel and Baldwin, 1991; 1992a; 1992b).

LuxR Protein

According to the model outlined above, LuxR protein binds VAI-1 and activates transcription from the *lux* operon promoter. The role for LuxR as an autoinducer-binding transcriptional activator is based in part on the requirement for LuxR in *E. coli* containing the *lux* operon and in *V. fischeri* to produce light in response to VAI-1 (Engebrecht *et al.*, 1983; Engebrecht and Silverman, 1984; Dunlap and Kuo, 1992). Consistent with this role also is the presence of a *lux* box (described above) in the *lux* operon promoter, as well as genetic and physiological evidence that LuxR binds VAI-1 (Shadel *et al.*, 1990; Slock *et al.*, 1990; Adar *et al.*, 1992; Adar and Ulitzur, 1993; Gray *et al.*, 1994; Hanzelka and Greenberg, 1995). However, the mechanism by which LuxR activates *lux* operon transcription has not been examined in detail until recently. Furthermore, it is theoretically possible that LuxR/VAI-1 operates not directly to activate transcription but indirectly. For example, LuxR/VAI-1 could function by relieving a repression of *lux* operon expression by LexA or another repressor protein (*e.g.,* Ulitzur, 1989; Ulitzur and Dunlap, 1995); the *lux* operon *lux* box exhibits substantial similarity to the LexA protein binding site found in *E. coli* promoters controlled by the SOS response (Ulitzur and Kuhn, 1988; Baldwin *et al.*, 1989). For these reasons and because *V. fischeri* LuxR serves as a model for new members of the emerging LuxR family of proteins (Fuqua *et al.*, 1996; Dunlap, 1997), substantial emphasis has been placed recently on understanding the structure and function of LuxR.

The *V. fischeri* LuxR protein, consisting of 250 amino acid residues, apparently associates with the cytoplasmic membrane (Engebrecht and Silverman, 1984; Kolibachuk and Greenberg, 1993). Synthesis of active LuxR and the binding of VAI-1 by the amino-terminal portion of LuxR (see below) are enhanced by GroEL (Figure 1), indicating that chaperonin-mediated folding is necessary to stabilize the protein in active form (Adar *et al.*, 1992; Dolan and Greenberg, 1992; Adar and Ulitzur, 1993; Hanzelka and Greenberg, 1995). The carboxy-terminal portion of LuxR (see below) exhibits remarkable thermal stability *in vitro*, regaining full activity after being heated to 70 °C for 30 min; either the protein is not denatured when heated or it folds back, upon cooling, into active form (Stevens and Greenberg, 1997) in the absence of GroEL.

Detailed analysis of full length LuxR has proven difficult; the overexpressed protein forms highly insoluble inclusion bodies (Kaplan and Greenberg, 1987). Deletion and mutational analyses of LuxR, however, have substantially overcome that impass. LuxR is composed of two modules or domains, an amino-terminal regulatory domain and a carboxy-terminal transcriptional activating domain. The amino-terminal domain, which binds VAI-1, regulates the activity of the carboxy-terminal domain, which associates with DNA and activates *lux* operon transcription. In the absence of VAI-1, the amino-terminal domain blocks the ability of the carboxy-terminal domain to associate with DNA. Binding of VAI-1 by the amino-terminal domain is interpreted as altering that interaction, permitting the carboxy-terminal domain then to bind to the *lux* regulatory region and activate *lux* operon transcription (Slock *et al.*, 1990; Shadel *et al.*, 1990; Choi and Greenberg, 1991; 1992a; Stevens *et al.*, 1994; Hanzelka and Greenberg, 1995; Egland and Greenberg, 1999). Alternatively, the amino-terminal domain might prevent a proposed transcriptionally functional multimerization of LuxR to occur. In the presence of VAI-1, that interaction would be altered, permitting a LuxR dimer then to form and associate with the dyadically symmetric *lux* box (Choi and Greenberg, 1992b; Fuqua *et al.*, 1996). Regardless, DNase I protection analysis and *in vitro* transcription assays with purified *lux* regulatory region DNA and purified proteins have demonstrated that the LuxR carboxy-terminal domain polypeptide and RNA polymerase bind synergistically to the *lux* box-*lux* operon promoter region and that the binding of these two proteins in this region is sufficient for *lux* operon mRNA synthesis (Stevens *et al.*, 1994; Stevens and Greenberg, 1997). The *in vitro* nature of these results with purified DNA and proteins argues against the hypothesis for a repressor protein that blocks *lux* operon transcription, the activity of which is alleviated by LuxR/VAI-1, and it argues against an hypothesized indirect effect of autoinducer, one in which autoinducer is an effector for another protein that activates LuxR (Sitnikov *et al.*, 1995). Instead, these results support a more simple model, one in which a mutually dependent interaction of the carboxy-terminal domain of LuxR, RNA polymerase, the *lux* box and the *lux* operon promoter region facilitates RNA polymerase binding to and initiating transcription from the *lux* operon promoter (Stevens and Greenberg, 1997).

Homologs of the *V. fischeri* LuxR protein have been identified in several other species of bacteria that use quorum sensing. The amino acid residue identities to *V. fischeri* LuxR range from 18% to 38% in these other species (Ochsner *et al.*, 1994; Fuqua *et al.*, 1996; Dunlap, 1997; Flavier *et al.*, 1997; Milton *et al.*, 1997; Puskas *et al.*, 1997; Swift *et al.*, 1997; Nasser *et al.*, 1998; Lewenza *et al.*, 1999). The similarities between the LuxR homologs tend to occur in the amino-terminal domain and in the helix-turn-helix portion of the C-terminal domain (Fuqua *et al.*, 1994). The C-terminal domain of LuxR exhibits similarity to the DNA-binding region of members of the UphA/FixJ family of proteins (Fuqua *et al.*, 1994; Fuqua *et al.*, 1996). Furthermore, the deduced translation products of *luxR* genes from different strains of *V. fischeri* exhibit greater than 80% identity (Gray and Greenberg, 1992).

LuxI Protein and Synthesis of VAI-1

The first gene of the *lux* operon, *luxI*, directs the synthesis of a 25 kDa protein that is necessary and sufficient for *E. coli* and *V. fischeri* to produce VAI-1 (Engebrecht *et al.*, 1983; Engebrecht and Silverman, 1984; Kuo *et al.*, 1994; 1996). Questions central to understanding quorum sensing in *V. fischeri* have included whether LuxI is directly responsible for synthesis of VAI-1, and, if so, whether LuxI catalyzes VAI-1 synthesis from 3-oxo-hexanoyl-CoA and *S*-adenosylmethionine (SAM) (Eberhard *et al.*, 1991; Hanzelka and Greenberg, 1996), or possibly from other substrates.

Recently, *in vitro* analysis of LuxI resolved these questions. Schaefer *et al.* (1996), relying on the finding that in *V. fischeri*, *luxI*, in addition to directing the synthesis of VAI-1, also directs the synthesis of hexanoyl-HSL (VAI-3) (Kuo *et al.*, 1994), for which the probable substrates were more readily available than for VAI-1, demonstrated that a purified maltose binding protein-LuxI fusion protein catalyzes the synthesis of VAI-3 from hexanoyl-acyl carrier protein (hexanoyl-ACP) and SAM. No additional factor and no energy source was required for the activity. These results, in demonstrating the acyl-HSL synthase activity of LuxI, established that LuxI functions directly in autoinducer synthesis (Schaefer *et al.*, 1996) and not through other cellular enzymes as had been proposed (Salmond *et al.*, 1995). Acyl-HSL synthase activity has also been demonstrated for the product of *traI*, a gene necessary for synthesis of autoinducer in *A. tumefaciens*; purified hexahistidinyl-TraI protein utilized 3-oxo-octanoyl-ACP and SAM to synthesize 3-oxo-octanoyl-HSL (More *et al.*, 1996). Also, the enzymatic activity of purified *P. aeruginosa* RhlI, using SAM and either butanoyl-CoA or hexanoyl-CoA to produce the corresponding acyl-HSLs, was demonstrated recently (Jiang *et al.*, 1998).

Several proteins with sequence similarity to LuxI (25 to 35% identity) have been identified in various other species of bacteria producing acyl-HSLs (Fuqua *et al.*, 1996; Dunlap, 1997; Milton *et al.*, 1997; Swift *et al.*, 1997; Flavier *et al.*, 1997; Nasser *et al.*, 1998; Lewenza *et al.*, 1999). Also, the deduced LuxI sequences of different *V. fischeri* strains exhibit a high degree of identity (Gray and Greenberg, 1992). A recent mutational analysis defined two regions of LuxI important for its activity as an acyl-HSL synthase, amino acid residues 25-70, proposed as the active site, and 104-164, possibly involved in substrate specificity; these regions contain many of the residues conserved in the members of LuxI family of proteins (Hanzelka *et al.*, 1997).

Heirarchical Control of Quorum Sensing in *V. fischeri*

The LuxI/LuxR quorum sensing mechanism in *V. fischeri* is controlled by the intracellular signal cAMP and CRP (Figure 1). cAMP/CRP are required for transcription of *luxR*. The dominance of this control over the quorum sensing mechanism is revealed by the absence of luminescence induction in *cya*-like and *crp*-like mutants of *V. fischeri* (Dunlap, 1989). Early studies identified a transient repression of luminescence by glucose (Ruby and Nealson, 1976; Friedrich and Greenberg, 1983) that was suggestive of cAMP control of *lux* gene expression. Isolation of the *V. fischeri lux* genes (Engebrecht *et al.*, 1983) then made obvious and feasible the use of *E. coli cya* and *crp* mutants with which to examine that possibility. CRP and cAMP were shown to be necessary for induction of luminescence, first in *E. coli* and later in *V. fischeri*, and to function by activating transcription from the *luxR* promoter (Dunlap and Greenberg, 1985; 1988; Dunlap and Ray, 1989; Dunlap, 1989; Dunlap and Kuo, 1992). Activation of *luxR* transcription apparently potentiates the quorum sensing response by building up the cellular level of LuxR protein. A consensus cAMP/CRP-binding site is present in the *luxR* promoter region (Devine *et al.*, 1988; Engebrecht and Silverman, 1987), and CRP has been shown to bind to the region encompassing this site (Shadel *et al.*, 1990; Stevens *et al.*, 1994; Stevens and Greenberg, 1997). Additional regulatory complexity is indicated by the 2 to 10-fold repression of *lux* operon transcription by cAMP/CRP, which mirrors the repression of *luxR* expression by VAI-1/LuxR (Dunlap and Greenberg, 1985; Dunlap and Greenberg, 1988; Dunlap and Kuo, 1992).

Control of quorum regulation by cAMP/CRP might serve to funnel information on the nutritional and growth rate status of the cell into the quorum sensing mechanism. The dominance of cAMP/CRP control suggests that it functions to stimulate the quorum sensing response under conditions of nutritional and growth rate limitation, when cAMP levels in *V. fischeri* cells presumably are high, and to delay it or block it completely under conditions of rapid growth, when cellular levels of cAMP presumably would be low.

Various culture conditions influence growth and luminescence in *V. fischeri* in opposite ways, enhancing one while suppressing the other, including glucose, oxygen and iron (Ruby and Nealson, 1976; Nealson and Hastings, 1977; Haygood and Nealson, 1985a; 1985b; Hastings *et al.*, 1987), but how those conditions are coupled to *lux* operon induction is not yet understood (Dunlap, 1992a; 1992b). If those conditions influence the cellular level of cAMP, for example, that influence, via its effect on cAMP/CRP control of *luxR* expression, would provide a mechanistic link.

Other quorum sensing systems apparently also are subject to a dominant control by cAMP/CRP. In *P. aeruginosa*, *lasR* encodes a LuxR homolog that is at the top of a quorum sensing heirarchy in that species; expression of *lasR* is controlled by Vfr, a CRP homolog, from a consensus cAMP/CRP binding site in the *lasR* promoter (Albus *et al.*, 1997). Similarly in *Erwinia chrysanthemi*, expression of the LuxR homolog *expR* is positively controlled by CRP from a cAMP/CRP binding site in the *expR* promoter region (Nasser *et al.*, 1998; Reverchon *et al.*, 1998).

A Second Quorum Sensing System in *V. fischeri*

The quorum sensing system of *V. fischeri* contains a second autoinducer synthase gene, *ainS*, which directs the synthesis of a second autoinducer, octanoyl-HSL (VAI-2) (Figure 1). VAI-2 via LuxR negatively modulates *lux* operon transcription. Evidence for the second quorum sensing system developed from construction of a non-polar *luxI* deletion mutant of *V. fischeri*, MJ-211. Despite the mutation, which eliminated the ability of cells to produce VAI-1 and which was expected to result in a non-luminous phenotype, the mutant produced a low but detectable level of light. Furthermore, light production during growth of the mutant in batch culture showed the same pattern of decline and rise back as the parental strain, a pattern characteristic of quorum regulation. The luminescence behavior of the mutant suggested that *V. fischeri* cells produce a *luxI*-independent inducing factor. Purification and identification of that factor as octanoyl-HSL (VAI-2) confirmed the presence of the second quorum sensing system, as did isolation of a fragment of *V. fischeri* DNA distinct from *luxI* that was necessary and sufficient for *E. coli* to produce VAI-2 (Kuo *et al.*, 1994).

Genetic analysis of that fragment revealed the presence of two genes, *ainS*, which is required for production of VAI-2, and *ainR*, which might encode a regulatory protein (Gilson *et al.*, 1995). The deduced amino acid sequence of *ainS* exhibited no significant similarity to LuxI or other members of the LuxI family of autoinducer synthesis proteins (Gilson *et al.*, 1995), now defined enzymatically as acyl-HSL synthases (Schaefer *et al.*, 1996; More *et al.*, 1996). The C-terminal half of AinS is homologous (34% identity), however, to the *V. harveyi* LuxM protein (Gilson *et al.*, 1995). LuxM and the upstream gene LuxL have been implicated in synthesis of a *V. harveyi* quorum sensing signal for luminescence (Bassler *et al.*, 1993). The AinS-LuxM homology led Gilson *et al.* (1995) to propose that AinS and LuxM are members of a new family of autoinducer synthesis proteins. Open issues, however, have been whether AinS exhibits enzymatic activity and if so, whether it uses substrates similar to those used by the LuxI family of acyl-HSL synthases. A recent study has resolved those issues by demonstrating *in vitro* the acyl-HSL synthase activity of a purified maltose binding protein-AinS fusion protein, which uses SAM and octanoyl-ACP to produce octanoyl-HSL (Hanzelka *et al.*, 1999). Furthermore, recent evidence obtained through 2-dimensional polyacrylamide gel electrophoresis (2-D PAGE) analysis of quorum sensing mutants indicates that VAI-2 inhibits production of Lux proteins and the production of the several newly identified non-Lux proteins composing a quorum sensing regulon in *V. fischeri* (Callahan and Dunlap, manuscript in preparation).

VAI-2 can operate *in vivo* as a negative modulator of *lux* operon transcription. By itself, VAI-2 stimulates luminescence a small amount and does so in a LuxR-dependent manner (Eberhard *et al.*, 1986; Kuo *et al.*, 1994), indicating the ability of VAI-2 via LuxR to weakly activate *lux* operon transcription (Figure 1). Furthermore, VAI-2 interferes with the luminescence stimulating activity of VAI-1, suggesting a competitive inhibition by VAI-2 of the interaction between VAI-1 and LuxR (Eberhard *et al.*, 1986; Kuo *et al.*, 1996). Consistent with these observations, a *V. fischeri* mutant defective in *ainS* induces luminescence at a lower population density and more rapidly than the parent strain (Kuo *et al.*, 1996). The inhibitory activity of VAI-2 has been hypothesized to help prevent premature induction of luminescence, i.e., at low population densities (Kuo *et al.*, 1996).

Various aspects of VAI-2 and *ainSR* remain unresolved at this time (Figure 1). Perhaps most intriguing is whether VAI-2 via LuxR, or possible via AinR, regulates the expression of genes other than *lux* (Kuo *et al.*, 1994). As described below, LuxR and VAI-1 have been found recently to activate the production of at least seven non-Lux proteins in *V. fischeri*, but so far no protein whose production is dependent on VAI-2 has been identified (Callahan and Dunlap, manuscript in preparation). A regulatory role for AinR also has not been identified, but one seems possible given the partial identity of AinR with the amino terminus of the *V. harveyi luxN* gene product, which has been proposed to function as a sensor/receptor for *V. harveyi* autoinducer (Bassler *et al.*, 1993). How *ainSR* transcription is controlled also remains undefined; however, the *ainSR* regulatory region contains a *lux* box, which implies transcriptional control by LuxR and either VAI-1 or VAI-2 (Gilson *et al.*, 1995).

Quorum Regulation of Non-*lux* Genes in *V. fischeri*

Previously, luminescence was the only activity known to be regulated in *V. fischeri* by LuxR and acyl-HSLs. Recent studies have identified, however, several *V. fischeri* genes other than those of the *lux* operon that are quorum regulated. An important step in identifying these genes was the construction of a mutant entirely defective in the synthesis of *V. fischeri* autoinducers. The identification of VAI-2 and the cloning of the *ainS* gene made this step both necessary, in demonstrating the presence of a second signal, one with demonstrated cross activity on *lux* operon transcription, and obvious, by providing the genetic locus for that activity. A double acyl-HSL synthase mutant, MJ-215, defective in both *luxI* and in *ainS*, was constructed by replacement of the wild-type *ainS* gene with a mutated form in the *luxI* deletion background. The double mutant does not induce luminescence and makes neither VAI-1 nor VAI-2; it responds, however, to exogenously added VAI-1 with the production of a high level of light (Kuo *et al.*, 1996), demonstrating that the LuxR and luminescence functions are intact.

Two complementary approaches have been used with this strain to identify quorum regulated non-*lux* genes in *V. fischeri*. The first approach, 2-dimensional polyacrylamide gel electrophoresis (2-D PAGE) of total cellular proteins, permitted visualization of proteins produced under the control of *V. fischeri* acyl-HSLs. With this method, five non-Lux proteins whose production is dependent on VAI-1 and LuxR, and their genes, have been identified (Callahan and Dunlap, manuscript in preparation). The second approach employed *lacZ*-fusion technology via transposon MudI, and that approach has led to the identification of two non-*lux*, quorum sensing regulated genes, both of which differ from those identified by 2-D PAGE analysis (Callahan and Dunlap, unpublished data). The identification of these seven genes reveals the presence of a LuxR/VAI-1-dependent quorum sensing regulon in *V. fischeri*; the luminescence system is one phenotypically distinct part of this regulon (Callahan and Dunlap, manuscript in preparation). Recent evidence indicates that the quorum sensing regulon is composed of additional genes (Callahan and Dunlap, unpublished data), which would be consistent with the large number of quorum-regulated genes identified in certain other bacteria (*e.g.*, Van Delden and Iglewski, 1998). Analysis of the functions of the proteins coordinately produced under quorum sensing control by *V. fischeri* may help define the nature of the environmental conditions important in this bacterium's adaptation to high population density and host association.

Perspective

Autoinduction of luminescence, thought at one time to be a peculiar attribute of a few obscure marine bacteria, is now recognized as a model quorum sensing system with wide relevance in basic and applied research on bacterial gene regulation and host association. One can wonder, from the perspective available today and with awareness of the years of dedicated effort by individuals and small groups during the 1960's, 1970's and 1980's in defining the mechanism of luminescence gene regulation, what other neglected or unknown bacterial activities, if given similar effort, would lead to new paradigms in bacterial physiology and genetics. Certainly, the long-neglected marine environment would appear to be a good hunting ground for such activities.

Acknowledgements

I thank S. M. Callahan and E. P. Greenberg for permission to cite unpublished information. Research in the author's laboratory is supported by NSF grant MCB 97-22972.

References

Albus, A.M., Pesci, E.C., Runyen-Janecky, L.J., West, S.E.H., and Iglewski, B.H. 1997. Vrf controls quorum sensing in *Pseudomonas aeruginosa*. J. Bacteriol. 179: 3928-3935.

Adar, Y.Y., Simaan, M., and Ulitzur, S. 1992. Formation of the LuxR protein in the *Vibrio fischeri lux* system is controlled by HtpR through the GroESL proteins. J. Bacteriol. 174: 7138-7143.

Adar, Y.Y., and Ulitzur, S. 1993. GroESL proteins facilitate binding of externally added inducer by LuxR protein-containing *E. coli* cells. J. Biolumin. Chemilumin. 8: 261-266.

Baldwin, T.O., Devine, J.H., Heckel, R.C., Lin, J.-W., Shadel, G.S. 1989. The complete nucleotide sequence of the *lux* regulon of *Vibrio fischeri* and the *luxABN* region of *Photobacterium leiognathi* and the mechanism of control of bacterial bioluminescence. J. Biolumin. Chemilumin. 4: 326-341.

Bassler, B.L., Wright, M., Showalter, R.E., and Silverman, M.R. 1993. Intercellular signalling in *Vibrio harveyi*, sequence and function of genes regulating expression of luminescence. Mol. Microbiol. 9: 773-786.

Baumann, P., and Baumann, L. 1981. The marine Gram-negative eubacteria: genera *Photobacterium*, *Beneckea*, *Alteromonas*, *Pseudomonas*, and *Alcaligenes*. In: The Prokaryotes. A Handbook on Habitats, Isolation, and Identification of Bacteria. M.P. Starr, H. Stolp, H.G. Trüper, A. Balows, H.G. and Schlegel, eds. Springer-Verlag, Berlin. p. 1302-1331.

Boettcher, K.J., and Ruby, E.G. 1995. Detection and quantification of *Vibrio fischeri* autoinducer from symbiotic squid light organs. J. Bacteriol. 177: 1053-1058.

Boylan, M., Graham, A.F., and Meighen, E.A. 1985. Functional identification of the fatty acid reductase components encoded in the luminescence operon of *Vibrio fischeri*. J. Bacteriol. 163: 1186-1190.

Boylan, M., Miyamoto, C., Wall, L., Graham, A.F., and Meighen, E.A. 1989. Lux C, D and E genes of the *Vibrio fischeri* luminescence operon code for the reductase, transferase, and synthetase enzymes involved in aldehyde biosynthesis. Photochem. Photobiol. 49: 681-688.

Choi, S.H., and Greenberg, E.P. 1991. The C-terminal region of the *Vibrio fischeri* LuxR protein contains an inducer-independent *lux* gene activating domain. Proc. Natl. Acad. Sci. USA. 88: 11115-11119.

Choi, S.H., and Greenberg, E.P. 1992a. Genetic dissection of DNA binding and luminescence gene activation by the *Vibrio fischeri* LuxR protein. J. Bacteriol. 174: 4064-4069.

Choi, S.H., and Greenberg, E.P. 1992b. Genetic evidence for multimerization of LuxR, the transcriptional activator of *Vibrio fischeri* luminescence. Mol. Mar. Biol. Biotechnol. 1: 408-413.

Claes, M.F., and Dunlap, P. V. 1999. Aposymbiotic culture of the sepiolid squid *Euprymna scolopes*: role of the symbiotic bacterium *Vibrio fischeri* in host animal growth, development, and light organ morphogenesis. J. Exp. Zool. 286: 000-000 (in press).

Devine, J.H., Countryman, C., and Baldwin, T.O. 1988. Nucleotide sequence of the *luxR* and *luxI* genes and structure of the primary regulatory region of the *lux* regulon of *Vibrio fischeri* ATCC 7744. Biochem. 27: 837-842.

Devine, J.H., Shadel, G.S., and Baldwin, T.O. 1989. Identification of the operator of the *lux* regulon from the *Vibrio fischeri* strain ATCC 7744. Proc. Natl. Acad. Sci. USA. 86: 5688-5692.

Dolan, K.M., and Greenberg, E.P. 1992. Evidence that GroEL, not σ^{32}, is involved in transcription regulation of the *Vibrio fischeri* luminescence genes in *Escherichia coli*. J. Bacteriol. 174: 5132-5135.

Dunlap, P.V. 1985. Osmotic control of luminescence and growth in *Photobacterium leiognathi* from ponyfish light organs. Arch. Microbiol. 141: 44-50.

Dunlap, P.V. 1989. Regulation of luminescence by cyclic AMP in *cya*-like and *crp*-like mutants of *Vibrio fischeri*. J. Bacteriol. 171: 1199-1202.

Dunlap, P.V. 1991. Organization and regulation of bacterial luminescence genes. Photochem. Photobiol. 54: 1157-1170.

Dunlap, P.V. 1992a. Iron control of the *Vibrio fischeri* luminescence system in *Escherichia coli*. Arch. Microbiol. 157: 235-241.

Dunlap, P.V. 1992b. Mechanism for iron control of the *Vibrio fischeri* luminescence system: involvement of cyclic AMP and cyclic AMP receptor protein and modulation of DNA level. J. Biolumin. Chemilumin. 7: 203-214.

Dunlap, P.V. 1997. *N*-Acyl-L-homoserine lactone autoinducers in bacteria: unity and diversity. In: Bacteria as Multicellular Organisms. J.A. Shapiro and J. Dworkin, eds. Oxford University Press, New York. p. 69-106.

Dunlap, P.V., and Greenberg, E.P. 1985. Control of *Vibrio fischeri* luminescence gene expression in *Escherichia coli* by cyclic AMP and cyclic AMP receptor protein. J. Bacteriol. 164: 45-50.

Dunlap, P.V., and Greenberg, E.P. 1988. Control of *Vibrio fischeri lux* gene transcription by a cyclic AMP receptor protein-LuxR protein regulatory circuit. J. Bacteriol. 170: 4040-4046.

Dunlap, P.V., and Greenberg, E.P. 1991. Role of intercellular chemical communication in the *Vibrio fischeri*-monocentrid fish symbiosis. In: Microbial Cell-Cell Interactions. M. Dworkin, ed. American Society for Microbiology, Washington, D.C. p. 219-253.

Dunlap, P.V., and Kuo, A. 1992. Cell density-dependent modulation of the *Vibrio fischeri* luminescence system in the absence of autoinducer and LuxR protein. J. Bacteriol. 174: 2440-2448.

Dunlap, P.V., and Ray, J.M. 1989. Requirement for autoinducer in transcriptional negative autoregulation of the *Vibrio fischeri luxR* gene in *Escherichia coli*. J. Bacteriol. 171: 3549-3552.

Eberhard, A. 1972. Inhibition and activation of bacterial luciferase synthesis. J. Bacteriol. 109: 1101-1105.

Eberhard, A., Burlingame, A.L., Eberhard, C., Kenyon, G.L., Nealson, K.H., and Oppenheimer, N.J. 1981. Structural identification of autoinducer of *Photobacterium fischeri* luciferase. Biochem. 20: 2444-2449.

Eberhard, A., Longin, T., C.A. Widrig, C.A., and Stranick, S.J. 1991. Synthesis of the *lux* gene autoinducer in *Vibrio fischeri* is positively autoregulated. Arch. Microbiol. 155: 294-297.

Eberhard, A., Widrig, C.A., McBath, P., and Schineller, J.B. 1986. Analogs of the autoinducer of bioluminescence in *Vibrio fischeri*. Arch. Microbiol. 146: 35-40.

Egland, K.A., and Greenberg, E.P. 1999. Quorum sensing in *Vibrio fischeri*:

elements of the *luxI* promoter. Mol. Microbiol. 31: 1197-1204.

Engebrecht, J., Nealson, K., and Silverman, M. 1983. Bacterial bioluminescence, isolation and genetic analysis of functions from *Vibrio fischeri*. Cell. 32: 773-781.

Engebrecht, J., and Silverman, M. 1984. Identification of genes and gene products necessary for bacterial bioluminescence. Proc. Natl. Acad. Sci. USA. 81: 4154-4158.

Engebrecht, J., and Silverman, M. 1986. Regulation and expression of bacterial genes for bioluminescence. Genet. Eng. 8: 31-44.

Engebrecht, J., and Silverman, M. 1987. Nucleotide sequence of the regulatory locus controlling expression of bacterial genes for bioluminescence. Nucleic Acids Res. 15: 10455-10467.

Flavier, A.B., Ganova-Raeva, L.M., Schell, M.A., and Denny, T.P. 1997. Hierarchical autoinduction in *Ralstonia solanacearum*: control of *N*-acyl-homoserine lactone production by a novel autoregulatory system resonsive to 3-hydroxypalmitic acid methyl ester. J. Bacteriol. 179: 7089-7097.

Friedrich, W.F., and Greenberg, E.P. 1983. Glucose repression of luminescence and luciferase in *Vibrio fischeri*. Arch. Microbiol. 134: 87-91.

Fuqua, W.C., and Winans, S.C. 1994. A LuxR-LuxI type regulatory system activates *Agrobacterium* Ti plasmid conjugal transfer in the presence of a plant tumor metabolite. J. Bacteriol. 176: 2796-2806.

Fuqua, W.C., Winans, S.C., and Greenberg, E.P. 1994. Quorum sensing in bacteria, the LuxR-LuxI family of cell density-responsive transcriptional regulators. J. Bacteriol. 176: 269-275.

Fuqua, W.C., Winans, S.C., and Greenberg, E.P. 1996. Census and consensus in bacteiral ecosystems: the LuxR-LuxI family of quorum-sensing transcriptional regulators. Annu. Rev. Microbiol. 50: 727-751.

Gilson, L., Kuo, A., and Dunlap, P. V. 1995. AinS and a new family of autoinducer synthesis proteins. J. Bacteriol. 177: 6946-6951.

Gomer, R.H. 1999. Cell density sensing in a eukaryote. ASM News. 65: 23-29.

Gray, K.M., and Greenberg E.P. 1992. Sequencing and analysis of *luxR* and *luxI*, the luminescence regulatory genes from the squid light organ symbiont *Vibrio fischeri* ES114. Mol. Mar. Biol. and Biotechnol. 1: 414-419.

Gray, K.M., Passador, L., Iglewski, B.H., and Greenberg, E.P. 1994. Interchangeability and specificity of components from the quorum-sensing regulatory systems of *Vibrio fischeri* and *Pseudomonas aeruginosa*. J. Bacteriol. 176: 3076-3080.

Greenberg, E.P. 1997. Quorum sensing in Gram-negative bacteria. ASM News. 63: 371-377.

Hanzelka, B.L., and Greenberg, E.P. 1995. Evidence that the N-terminal region of the *Vibrio fischeri* LuxR protein constitutes an autoinducer-binding domain. J. Bacteriol. 177: 815-817.

Hanzelka, B.L., and Greenberg, E.P. 1996. Quorum sensing in *Vibrio fischeri*: evidence that *S*-adenosylmethionine is the amino acid substrate for autoinducer synthesis. J. Bacteriol. 178: 5291-5294.

Hanzelka, B.L., Stevens, A.M., Parsek, M.R., Crone, T.J., and Greenberg, E.P. 1997. Mutational analysis of the *Vibrio fischeri* LuxI polypeptide: critical regions of an autoinducer synthase. J. Bacteriol. 179: 4882-4887.

Hanzelka, B.L., Parsek, M.R., Val, D.L., Dunlap, P.V. Cronan, J.E., and Greenberg, E.P. 1999. Acylhomoserine lactone synthase activity of the *Vibrio fischeri* AinS protein. J. Bacteriol. 181: 5766-5770.

Hastings, J.W., and Greenberg, E.P. 1999. Quorum sensing: the explanation of a curious phenomenon reveals a common characteristic of bacteria. J. Bacteriol. 181: 2667-2668.

Hastings, J.W., and Nealson, K.H. 1981. The symbiotic luminous bacteria. In: The Prokaryotes. M.P. Starr, H. Stolp, H.G. Truper, A. Baows, and H.G. Schlegel, eds. Springer, New York, N.Y. p.1332-1345.

Hastings, J.W., Makemson, J.C., and Dunlap, P.V. 1987. How are growth and luminescence regulated independently in light organ symbiosis? Symbiosis. 4: 3-24.

Haygood, M.G., and Nealson, K.H. 1985a. The effect of iron on the growth and luminescence of the symbiotic bacterium *Vibrio fischeri*. Symbiosis. 1: 39-51.

Haygood, M.G., and Nealson, K.H. 1985b. Mechanism of iron regulation of luminescence in *Vibrio fischeri*. J. Bacteriol. 162: 209-216.

Hwang, I., Li, P., Zhang, L., Piper, K.R., Cook, D.M., Tate, M.E., and Farrand, S.K. 1994. TraI, a LuxI homologue, is responsible for production of conjugation factor, the Ti plasmid *N*-acylhomoserine lactone autonducer. Proc. Natl. Acad. Sci. USA. 91: 4639-4643.

Jiang, Y., Camara, M., Chhabra, S.R., Hardie, K.R., Bycroft, B.W., Lazdunski, A., Salmond, G.P.C., Stewart, G.S.A.B., and Williams, P. 1998. *In vitro* biosynthesis of the *Pseudomonas aeruginosa* quorum-sensing signal molecule *N*-butanoyl-L-homoserine lactone. Mol. Microbiol. 28: 193-203.

Kaplan, H.B., and Greenberg, E.P. 1985. Diffusion of autoinducer is involved in regulation of the *Vibrio fischeri* luminescence system. J. Bacteriol. 163: 1210-1214.

Kaplan, H.B., and Greenberg, E.P. 1987. Overproduction and purification of the *luxR* gene product, the transcriptional activator of the *Vibrio fischeri* luminescence system. Proc. Natl. Acad. Sci. USA. 84: 6639-6643.

Kleerebezem, M., Quadri, L.E.N., Kuipers, O.P., and de Vos, W.M. 1998. Quorum sensing by peptide pheromones and two-component signal-transduction systems in Gram-positive bacteria. Mol. Microbiol. 24: 895-904.

Kolibachuk, D., and Greenberg, E.P. 1993. The *Vibrio fischeri* luminescence gene activator LuxR is a membrane-associated protein. J. Bacteriol. 175: 7307-7312.

Kuo, A., Blough, N.V., and Dunlap, P.V. 1994. Multiple *N*-acyl-homoserine lactone autoinducers of luminescence in the marine symbiotic bacterium *Vibrio fischeri*. J. Bacteriol. 176: 7558-7565.

Kuo, A., Callahan, S.M., and Dunlap, P.V. 1996. Modulation of luminescence operon expression by *N*-octanoyl-homoserine lactone in *ainS* mutants of *Vibrio fischeri*. J. Bacteriol. 178: 971-976.

Latifi, A., Winson, M.K., Foglino, M., Bycroft, B.W., Stewart, G.S.A.B., Lazdunski, A., and Williams, P. 1995. Multiple homologues of LuxR and LuxI control expression of virulence determinants and secondary metabolites through quorum sensing in *Pseudomonas aeruginosa* PAO1. Mol. Microbiol. 17: 333-343.

Lee, K.-H., and Ruby, E.G. 1992. The detection of the squid light organ symbiont *Vibrio fischeri* in Hawaiian seawater by using *lux* gene probes. Appl. Environ. Microbiol. 58: 942-947.

Lewenza, S., Conway, B., Greenberg, E.P., and Sokol, P.A. 1999. Quorum sensing in *Burkholderia cepacia*: identification of the LuxRI homologs CepRI. J. Bacteriol. 181: 748-756.

McFall-Ngai, M.J. 1990. Crypsis in the pelagic environment. Am. Zool. 30: 175-188.

McFall-Ngai, M.J., and Dunlap, P.V. 1983. Three new modes of luminescence in the leiognathid fish *Gazza minuta* (Percifomes: Leiognathidae): discrete projected luminescence, ventral body flash and buccal luminescence. Mar. Biol. 73: 227-237.

Meighen, E.A., and Dunlap, P.V. 1993. Physiological, biochemical and genetic control of bacterial bioluminescence. Adv. Microbial Physiol. 34: 1-67.

Milton, D.L., Hardman, A., Camara, M., Chhabra, S.R., Bycroft, B.W., Stewart, G.S.A.B., and Williams, P. 1997. Quorum sensing in *Vibrio anguillarum*: characterization of the *vanI*/*vanR* locus and identification of the autoinducer *N*-(-oxodecanoyl)-L-homoserine lactone. J. Bacteriol. 179: 3004-3012.

More, M.I., Finger, L.D., Stryker, J.L., Fuqua, C., Eberhard, A., and Winans, S.C. 1996. Enzymatic synthesis of a quorum-sensing autoinducer through use of defined substrates. Science. 272: 1655-1658.

Nasser, W., Bouillant, M.L., Salmond, G., and Reverchon, S. 1998. Characterization of the *Erwinia chrysanthemi expI-expR* locus directing the synthesis of two *N*-acyl-homoserine lactone signal molecules. Mol. Microbiol. 29: 1391-1405.

Nealson, K. H. 1977. Autoinduction of bacterial luciferase. Occurrence, mechanism and significance. Arch. Microbiol. 112: 73-79.

Nealson, K.H., and Hastings, J.W. 1977. Low oxygen is optimal for luciferase synthesis in some bacteria. Ecological implications. Arch. Microbiol. 112: 9-16.

Nealson, K.H., and Hastings, J.W. 1979. Bacterial bioluminescence: its control and ecological significance. Microbiol. Rev. 43: 496-518.

Nealson, K.H., Platt, T., and Hastings, J.W. 1970. Cellular control of the synthesis and activity of the bacterial luminescent system. J. Bacteriol. 104: 313-322.

Ochsner, U.A., Koch, A.K., Fiechter, A., and Reiser, J. 1994. Isolation and characterization of a regulatory gene affecting rhamnolipid biosurfactant synthesis in *Pseudomonas aeruginosa*. J. Bacteriol. 176: 2044-2054.

Puskas, A., Greenberg, E.P., Kaplan, S., and Schaefer, A.L. 1997. A quorum-sensing system in the free-living photosynthetic bacterium *Rhodobacter sphaeroides*. J. Bacteriol. 179: 7530-7537.

Reverchon, S., Bouillant, M.L., Salmond, G., and Nasser, W. 1998. Integration of the quorum-sensing system in the regulatory networks controlling virulence factor synthesis in *Erwinia chrysanthemi*. Mol. Microbiol. 29: 1407-1418.

Rosson, R.A., and Nealson, K.H. 1981. Autoinduction of bacterialbioluminescence in a carbon-limited chemostat. Arch. Microbiol. 129: 299-304.

Ruby, E.G. 1996. Lessons from a cooperative bacterial-animal association:

the *Vibrio fischeri* - *Euprymna scolopes* light organ symbiosis. Annu. Rev. Microbiol. 50: 591-624.

Ruby, E.G., and Asato, L.M. 1993. Growth and flagellation of *Vibrio fischeri* during initiation of the sepiolid squid light organ symbiosis. Arch. Microbiol. 159: 160-167.

Ruby, E.G., and Nealson, K.H. 1976. Symbiotic association of *Photobacterium fischeri* with the marine luminous fish *Monocentris japonica*, a model of symbiosis based on bacterial studies. Biol. Bull. 141: 574-5867.

Ruby, E.G., and Nealson, K.H. 1978. Seasonal changes in the species composition of luminous bacteria in nearshore seawater. Limnol. Oceanogr. 23: 530-533.

Ruby, E.G., Greenberg, E.P., and Hastings, J.W. 1980. Planktonic marine luminous bacteria: species distribution in the water column. Appl. Environ. Microbiol. 39: 302-306.

Salmond, G.P.C., Bycroft, B.W., Stewart, G.S.A.B., and Williams, P. 1995. The bacterial 'enigma': cracking the code of cell-cell communication. Mol. Microbiol. 16: 615-624.

Schaefer, A.L., Hanzelka, B. L, Eberhard, A., and Greenberg, E. P. 1996. Quorum sensing in *Vibrio fischeri*: probing autoinducer-LuxR interactions with autoinducer analogs. J. Bacteriol. 178:2897-2901.

Schaefer, A.L., Val, D.L., Hanzelka, B. L, Cronan, J.E., Jr., and Greenberg, E. P. 1996. Generation of cell-to-cell signals in quorum sensing: acyl homoserine lactone synthase activity of a purified *Vibrio fischeri* LuxI protein. Proc. Natl. Acad. Sci. USA. 93: 9505-9509.

Shadel, G. S., and Baldwin, T. O. 1991. The *Vibrio fischeri* LuxR protein is capable of bidirectional stimulation of transcription and both positive and negative regulation of the *luxR* gene. J. Bacteriol. 173: 568-574.

Shadel, G.S., and Baldwin, T.O. 1992a. Identification of a distantly located regulatory element in the *luxD* gene required for negative autoregulation of the *Vibrio fischeri luxR* gene. J. Biol. Chem. 267: 7690-7695.

Shadel, G.S., and Baldwin, T.O. 1992b. Positive autoregulation of the *Vibrio fischeri luxR* gene. J. Biol. Chem. 267: 7696-7702.

Shadel, G.S., Young, R., and Baldwin, T.O. 1990. Use of regulated cell lysis in a lethal genetic selection in *Escherichia coli*, identification of the autoinducer-binding region of the LuxR protein from *Vibrio fischeri* ATCC 7744. J. Bacteriol. 172: 3980-3987.

Singley, C.T. 1983. *Euprymna scolopes*. In: Cephalopod Life Cycles, Vol. 1, Species Accounts. P.R. Boyle, ed. Academic Press, London. p. 69-74.

Sitnikov, D.M., Scineller, J.B., and Baldwin, T.O. 1995. Transcriptional regulation of bioluminescence genes from *Vibrio fischeri*. Mol. Microbiol. 17: 801-812.

Slock, J., VanReit, D., Kolibachuk, D., and Greenberg, E.P. 1990. Critical regions of the *Vibrio fischeri* LuxR protein defined by mutational analysis. J. Bacteriol. 172: 3974-3979.

Stevens, A.M., Dolan, K.M., and Greenberg, E.P. 1994. Synergistic binding of the *Vibrio fischeri* transcriptional activator domain and RNA polymerase to the *lux* promoter region. Proc. Natl. Acad. Sci. USA. 91: 12619-12623.

Stevens, A.M., and Greenberg, E.P. 1997. Quorum sensing in *Vibrio fischeri*: essential elements for activation of the luminescence genes. J. Bacteriol.

179: 557-562.

Swartzman, E., Kapoor, S., Graham, A.F., and Meighen, E A. 1990. A new *Vibrio fischeri lux* gene precedes a bidirectional termination site for the *lux* operon. J. Bacteriol. 172: 6797-6802.

Swift, S., Karlyshev, A.V., Fish, L., Durant, E.L., Winson, M.K., Chhabra, S.R., Williams, P., Macintyre, S., and Stewart, G.S.A.B. 1997. Quorum sensing in *Aeromonas hydrophila* and *Aeromonas salmonicida*: indentification of the LuxRI homologs AhyRI and AsaRI and their cognate *N*-acylhomoserine lactone signal molecules. J. Bacteriol. 179: 5271-5281.

Ulitzur, S. 1989. The regulatory control of the bacterial luminescence system — a new view. J. Biolumin. Chemilumin. 4: 317-325.

Ulitzur, S., and Dunlap, P. V. 1995. Regulatory circuitry controlling luminescence autoinduction in *Vibrio fischeri*. Photochem. Photobiol. 62: 625-632.

Ulitzur, S., and Kuhn, J. 1988. The transcription of bacteriol luminsecence is regulated by sigma 32. J. Biolumin. Chemilumin. 2: 81-93.

Van Delden, C., and Iglewski, B.H. 1998. Cell-to-cell signaling and *Pseudomonas aeruginosa* infections. Emerg. Infect. Dis. 4: 551-560.

Winson, M. K., Camara, M., Briggs, Litifi, A., Folino, Chhabra, S.R., Daykin, M., Bally, M., Chapon, V., Salmond, G.P.C., Bycroft, B., Lazdunski, A., Stewart, G.S.A.B., and Williams, P. 1995. Multiple *N*-acyl-L-homoserine lactone signal molecules regulate production of virulence determinants and secondary metabolites in *Pseudomonas aeruginosa*. Proc. Natl. Acad. Sci. USA. 92: 9427-9431.

Zenno, S., and Saigo, K. 1994. Identification of the genes encoding NAD(P)H-flavin oxidoreductases that are similar in sequence to *Escherichia coli* Fre in four species of luminous bacteria: *Photobacterium luminescens*, *Vibrio fischeri*, *Vibrio harveyi*, and *Vibrio orientalis*. J. Bacteriol. 176: 3544-3551.

Zenno, S., Saigo, K., Kanoh, H., and Inouye, S. 1994. Identification of the gene encoding the major NAD(P)H-flavin oxidoreductase of the bioluminescent bacterium *Vibrio fischeri* ATCC 7744. J. Bacteriol. 176: 3536-3543.

3

The *Euprymna scolopes* - *Vibrio fischeri* Symbiosis: A Biomedical Model for the Study of Bacterial Colonization of Animal Tissue

Edward G. Ruby

Pacific Biomedical Research Center
University of Hawaii, Manoa, Hawaii

Abstract

The diversity of microorganisms found in the marine environment reflects the immense size, range of physical conditions and energy sources, and evolutionary age of the sea. Because associations with living animal tissue are an important and ancient part of the ecology of many microorganisms, it is not surprising that the study of marine symbioses (including both cooperative and pathogenic interactions) has produced numerous discoveries of biotechnological and biomedical significance. The association between the bioluminescent bacterium *Vibrio fischeri* and the sepiolid squid *Euprymna scolopes* has emerged as a productive model system for the investigation of the mechanisms by which cooperative bacteria initiate colonization of specific host tissues. The results of the last decade of research on this system have begun to reveal surprising similarities between this association and the pathogenic associations of disease-causing *Vibrio* species, including those of interest to human health and aquaculture. Studies of the biochemical and molecular events underlying the development of the squid-vibrio symbiosis can be expected to continue to increase our understanding of the factors controlling both benign and pathogenic bacterial associations.

Introduction

Two areas of biomedical significance have emerged during the past 20 years. The first of these is the rise in coastal areas across the world of bacterial diseases borne by marine and brackish-water bacteria in the genus *Vibrio*

(Blake, 1983; Colwell, 1996). The second is the growing awareness of the importance of cooperative bacterial colonization of animal tissues to the proper development and health of the host (McFall-Ngai, 1998a; Rook and Stanford, 1998). Progress in addressing questions in both of these areas has been made through the investigation of a symbiotic association involving *Euprymna scolopes*, a shallow-water squid, and *Vibrio fischeri*, a bioluminescence bacterium. This association has developed into a paradigm not only for the study of symbiosis in the marine environ-ment, but for the myriad of both cooperative and pathogenic bacteria-animal host interactions that characterize the biological world.

The principal interactions between bacteria and animal tissue occur along the apical surfaces of host epithelial cells. Typically, a diverse consortium of microorganisms and host cells forms a dynamic, cooperating unit that helps maintain a healthy microenvironment within the colonized tissues. Such associations are ubiquitous, exhibit species specificity and, with few exceptions (Breznak and Brune, 1994; Whittaker *et al.*, 1996; Hooper *et al.*, 1998), are poorly understood. A better understanding of the process by which bacteria and host cells exchange the signals that underlie these phenomena is an important research goal of biomedical research. Over the past decade the *E. scolopes-V. fischeri* association, a natural and experimentally facile model system, has been used to discover the nature of the mechanisms by which animals form beneficial associations with microorganisms, and to determine how these mechanisms are related to, but have diverged from, those that function in pathogenic infections. Initial studies concentrated on a description of the system, leading to an understanding of the light organ morphology in both the juvenile (Montgomery and McFall-Ngai, 1993; Weis *et al.*, 1993; Hanlon, *et al.*, 1997) and the adult (McFall-Ngai and Montgomery, 1990; Montgomery and McFall-Ngai, 1998) animal. Similarly, the pattern of growth of *V. fischeri* cells during the initiation of the association (Ruby and Asato, 1993), and the role of the bacteria in the induction of its subsequent development (Foster and McFall-Ngai, 1998; McFall-Ngai and Ruby, 1998) were described. In addition, unexpected phenomena were discovered, such as the pronounced and dynamic diel rhythm dominating the state of the association (Lee and Ruby, 1994a; Boettcher *et al.*, 1996), and the presence of the antibacterial enzyme haloperoxidase within the contents of the light organ crypts (Weis *et al.*, 1993).

This review describes the colonization of the squid light organ by *V. fischeri*, emphasizing those bacterial determinants that have homologs or analogs among pathogenic *Vibrio* species. What we have learned about the development and specificity of this process by an examination of the host-imposed obstacles or barriers to colonization is also outlined. The review concludes with a discussion of the possible roles current and future studies of the *E. scolopes-V. fischeri* association may play in biomedically important areas of research.

The Symbiotic Partners

Euprymna scolopes is a small, nocturnal, sepiolid squid (Berry, 1912) that is bioluminescent owing to the presence of bacterial symbionts contained within a complex, bilobed, light-emitting organ (Nesis, 1987; McFall-Ngai and Montgomery, 1990). Light generated by these bacteria allows the host squid to produce a controlled, diffuse, ventral glow as an antipredatory behavior (McFall-Ngai, 1990). The organ lies within the center of the squid's mantle cavity, where it is continually bathed with ambient seawater (Figure 1). Histological and ultrastructural analyses of the light organ have revealed an integrated set of tissues that serves both to support nutritionally a culture of a specific luminous bacterium, and to control and direct the bioluminescence they produce (Montgomery and McFall-Ngai, 1992; McFall-Ngai, 1994). The bacterial symbionts are housed extracellularly within epithelium-lined spaces, or crypts. The polarized epithelium bears a dense layer of complex, branched microvilli within which the symbionts are invested (Lamarcq and McFall-Ngai, 1998).

V. fischeri, the only bacterium found to naturally persist in the light organ of E. scolopes (McFall-Ngai and Ruby, 1991), is a common marine microbe that can be easily isolated either from this symbiosis (Boettcher and Ruby, 1990) or from light organ associations of other squids (Fidopiastis et al., 1998) or fishes (Nealson and Hastings, 1991), and from free-living populations in the environment (Ruby and Lee, 1998), and that can be readily grown on

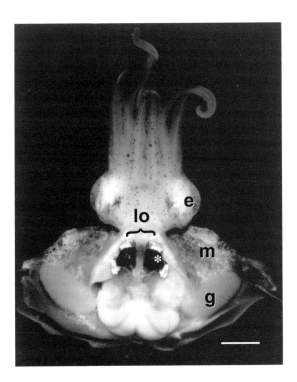

Figure 1. The Euprymna scolopes Symbiotic Light Organ
A ventral dissection of an anesthetized adult animal reveals the bilobed light organ (**lo**) in the center of the mantle cavity (McFall-Ngai and Montgomery, 1990), and partially occluded by the ink sac (black). Each lateral face of the adult light organ has a single pore through which the bacteria-containing crypts, deep in the center of the lobes of the organ (·), are connected to the outside environment (**e**, eye; **m**, mantle; **g**, gill; bar = 1 cm).

both complex and minimal media. An extensive literature exists concerning both the physiology and ecology of V. fischeri (Nealson and Hastings, 1991), and the biochemistry and molecular biology of its bioluminescence capability (Meighen, 1991; Dunlap, review in this issue).

Since 1996, new advances in the development of genetic and biochemical tools for the study of the V. fischeri-E. scolopes association have been important in facilitating progress in understanding this system. Principal among these bacterial genetic procedures and constructs for use in V. fischeri are:

(i)   Transfer of DNA by electroporation, using the ability to overcome innate restriction barriers in V. fischeri (Visick and Ruby, 1997),

(ii)   Chromosomal gene replacement and marker exchange (Visick and Ruby, 1996),

(iii)   Transposon mutagenesis using Tn10 (Visick and Ruby, 1997), and two mini-Tn5, carrying either a chloramphenicol (Graf and Ruby, 1994) or a streptomycin/spectinomycin (Stabb and Ruby, 1998) resistance cassette,

(iv)   Antibiotic resistance markers for erythromycin, tetracycline, trimethoprim (from Tn7) and streptomycin/spectinomycin (Visick and Ruby, 1997; Stabb and Ruby, 1998),

(v)   Stably maintained and suicide plasmids (Visick and Ruby, 1996), transferable by triparental mating (Stabb and Ruby, 1998),

(vi)   Promoter reporter constructs driving either lacZ or luxAB genes(Visick and Ruby, 1997; 1998a),

(vii) Libraries of V. fischeri DNA restricted with EcoRI (Visick and Ruby, 1998b), BglII or XbaI (Stabb and Ruby, 1998), and

(viii) A genetic database of over 60 V. fischeri genes and ORFs that has given rise to a reliable codon-usage table (E.G. Ruby, unpublished).

Recently developed molecular genetic tools have been used to demonstrate that a number of determinants, including motility (Graf et al., 1994), iron acquisition (Graf and Ruby, 1994), amino acid prototrophy (Graf and Ruby, 1998), luminescence (Visick and Ruby, 1996), and a periplasmic catalase (Visick and Ruby, 1998b), are required by V. fischeri for normal symbiotic competence (Table 1).

Similarly, valuable new methods have been devised for the study of the interaction of bacteria with specific E. scolopes tissues and cell types, including:

(i)   Development of a procedure by which to collect the contents of the light organ crypts (Graf and Ruby, 1996; Nyholm and McFall-Ngai, 1998),

(ii)   Description of bacteria-induced changes in host cell morphology (Foster and McFall-Ngai, 1998; McFall-Ngai, 1998b),

(iii) Identification of molecular markers for studies of population genetics (Nishiguchi et al., 1998), and

(iv) Construction of green-fluorescent-protein (GFP) expression vectors to label and visualize (by confocal laser microscopy) V. fischeri and other Vibrio species in the living host (McFall-Ngai, 1998b).

Taken together, these tools and approaches have put the investigation of the squid-vibrio association in a position to make significant advances at levels ranging from population biology and ecology to cell biochemistry and molecular genetics.

The Biology of the Association

The juvenile squid emerges from its egg free of bacteria (*i.e.*, 'aposymbiotic'), and must obtain an inoculum of *V. fischeri* cells to colonize its rudimentary light organ, and thereby become bioluminescent (Wei and Young, 1989; Montgomery and McFall-Ngai, 1994). In nature, inoculation is achieved by the symbiosis-competent *V. fischeri* cells typically present in the ambient seawater (Lee and Ruby, 1994a; 1995). The beating of cilia that cover two pairs of lateral projections on the surface of the nascent light organ (McFall-Ngai and Ruby, 1991; Montgomery and McFall-Ngai, 1993) serves to bring bacteria in the seawater close to a series of external pores. These pores lead via ducts to the interior of three pairs of crypts inside the light organ. Within 12 hours a few *V. fischeri* cells have entered the crypts, and have proliferated to a population size of between 10^5 and 10^6 cells (Ruby and Asato, 1993).

As it grows and matures, the squid nourishes and maintains within the light organ a *V. fischeri* population that eventually reaches as many as 10^9 cells. This nonpathogenic infection is characterized by two remarkable traits: the symbiont population remains monospecific, resisting any apparent contamination from other bacteria in the external environment (Montgomery

Table 1. Examples of Colonization Determinants that Characterize Both Pathogenic and Benign Infections

Proposed Role	Pathogenic *Vibrio* species[a]		Non-pathogenic *V. fischeri*	
	Determinant	Colonization phenotype[b]	Determinant	Colonization phenotype
Colonization regulation	ToxR [c]	+	ToxR [d]	?
Secreted ADPr enzyme	CtxAB [c]	+	HvnA/B [e]	?
Mucus penetration	motility [f]	+	motility [g]	+
Prototrophic growth	HemA [h]	+	LysA [i]	+
Iron acquisition	IrgA [j]	+	GlnD [k]	+
Adhesion	OmpU [l]	?	OmpV [m]	+
Catalase	——	?	KatA [n]	+
Luminescence	LuxAB [o]	?	LuxAB [p]	+
Mannose-sensitive adhesion	MshA [q]	-	(MshA) [r]	+
Protease	Hap [s]	-	(Hap)	?
hap regulation	HapR [t]	?	HapR [u]	?

[a] This list is a compendium of characteristics attributed to *V. cholerae*, *V. vulnificus*, and/or *V. parahaemolyticus*.
[b] The absence of the determinant either is well correlated to a colonization defect (+), is not correlated to a defect (-), or is as yet of an unclear or controversial importance (?).
[c] Salyers and Whitt, 1994
[d] Reich and Schoolnik, 1994
[e] Reich and Schoolnik, 1996; Reich *et al.*, 1997
[f] Richardson, 1991
[g] Graf *et al.*, 1994
[h] Rijpkema *et al.*, 1992
[i] Graf and Ruby, 1998
[j] Tashima *et al.*, 1996
[k] Graf and Ruby, 1994; 1996
[l] Sperandio *et al.*, 1995
[m] Aeckersberg and Ruby, 1998
[n] Visick and Ruby, 1998b
[o] Palmer and Colwell, 1991
[p] Visick and Ruby, 1996; Visick *et al.*, 1996
[q] Jouravleva *et al.*, 1998
[r] McFall-Ngai *et al.*, 1998; Feliciano and Ruby, 1999
[s] Hase and Finklestein, 1993; Wu *et al.*, 1996
[t] Jobling and Holmes, 1997
[u] P.M. Fidopiastis and E.G. Ruby, unpublished

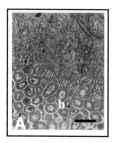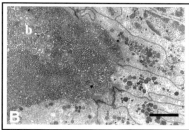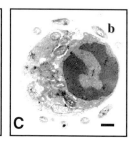

Figure 2. Transmission Electron Micrographic Images of Host Cell Types Encountered in the Light Organ by Symbiotic *Vibrio fischeri*
(A) Epithelial cells lining the light organ crypts bear dense microvilli in which *V. fischeri* cells (**b**) are embedded (bar = 2 μm). (B) Scattered throughout the epithelia, and near the bacteria, are goblet cells that contain a high density of internal vesicles (**v**), possibly containing glycoproteins (bar = 5 μm). (C) Phagocytic mollusk granulocytes are found within the crypts, moving among the bacterial population (bar = 1 μm).

and McFall-Ngai, 1993); and, there is no opportunistic invasion of other host tissues by these *V. fischeri* cells under normal circumstances (Ruby and McFall-Ngai, 1992; Ruby, 1996). These findings indicate the presence of a specific, tightly regulated recognition between the bacteria and the tissues of their host. In addition, population genetics studies (Lee and Ruby, 1994b) have shown that, while strains of *V. fischeri* isolated from different *Euprymna* species are capable of initiating and sustaining a symbiosis with *E. scolopes*, in mixed inoculations they are outcompeted by native symbiont strains (Nishiguchi *et al.*, 1998). Thus, the squid-vibrio association is both a complex and an ancient one, resulting from a coordinated evolution between the two partners.

Ultrastructural studies of both juvenile and adult light organs (McFall-Ngai and Montgomery, 1990; Montgomery and McFall-Ngai, 1993; Nyholm and McFall-Ngai, 1998) have revealed that *V. fischeri* associates with three types of host cells within the crypts - the highly polarized epithelium and specialized goblet cells that line the crypt space, and a population of amoeboid granulocytes (phagocytes) that moves freely within this space (Figure 2). These cell types all appear to participate in the dynamics of the colonization process. The *V. fischeri* cells that initiate the colonization must make their way through the ducts leading to the light organ interior, adhere to the epithelium of the crypts, and grow to fill the crypt spaces, utilizing host-secreted organic material. Every day at dawn, the squid expels most of the crypt contents, including 90 to 95% of the bacterial population, through the pores and into the external environment (Ruby and Asato, 1993; Lee and Ruby, 1994a); the remaining 5 to 10% of the bacteria continue to be associated with the microvillous epithelium. During the course of the day, these residual *V. fischeri* cells proliferate, repopulating the crypts and restoring a full bioluminescence potential for the host's nocturnal behavior (Boettcher *et al.*, 1996). This continuous cycle of purging and regrowth selects bacterial strains that not only are capable of continued growth within the crypts, but also are competitively dominant under the particular conditions cre-ated there (Lee and Ruby, 1994b; Visick and Ruby, 1998b).

Because the pores leading to the crypts remain open to the environment throughout the life of the squid, inappropriate species of bacteria that are present in the ambient seawater will continue to have access to the crypts (Lee and Ruby, 1994a). One line of defense against colonization by bacteria other than symbiosis-competent *V. fischeri* may be the phagocytic cells that are present in the crypts (Nyholm and McFall-Ngai, 1998). The function of these phagocytes is not well defined in the squid-vibrio system, but they are members of the class of cells that are the principal mediators of the immune response in squids (Budelmann *et al.*, 1997). These cells are thus candidates for a role in maintaining the light organ species-specific for *V. fischeri*, and/or in controlling the size of the symbiont population.

Cellular Events Occurring During Colonization

Detailed morphological, biochemical and genetic studies have revealed a number of dramatic events that accompany the initiation and persistence of the benign infection of the light organ crypts by *V. fischeri* cells. During the first few hours to days following the colonization of the crypt epithelium, both partners undergo significant symbiosis-related changes. For instance, the bacteria differentiate into a non-flagellated form (Ruby and Asato, 1993), and their level of luminescence per cell is induced over 1000-fold (Boettcher and Ruby, 1990) due to the accumulation in the crypts of at least one *V. fischeri*-specific quorum-sensing molecule, an acyl-homoserine lactone called VAI-1 (Fuqua *et al.*, 1996; Boettcher and Ruby, 1995).

The squid host undergoes an even more striking developmental program. Morphologically, before colonization is initiated, the nascent light organ has two pairs of ciliated projections on its surface (Montgomery and McFall-Ngai, 1993), and the polarized epithelial cells that line the crypts are columnar and have sparse microvillar projections (Lamarcq and McFall-Ngai, 1998). After *V. fischeri* cells have initiated their cooperative interaction, the gross morphology of these light organ structures begins to change. Within 8 hours following inoculation the ciliated projections begin to regress through a process of programmed cell death (Montgomery and McFall-Ngai, 1994; Foster and McFall-Ngai, 1998), and after a few days they have completely disappeared (Doino and McFall-Ngai, 1995). In addition, the epithelial cells lining the crypts have increased 4-fold in volume and become more cuboidal, and their microvilli have increased 4-fold in density and more completely surround and support the bacterial cells (Figure 2A) (McFall-Ngai, 1994; Lamarcq and McFall-Ngai, 1998). There are also significant biochemical changes during this period. Initially, a high concentration of a species of host mRNA encoding haloperoxidase (HPO) (Weis *et al.*, 1996) is present in the uncolonized light organ. However, once the organ is colonized, the level of HPO drops significantly (Small and McFall-Ngai, 1998), thus reducing the potential for the production of the antibacterial compound hypohalous acid (Forman and Thomas, 1986; Tomarev *et al.*, 1993; Boettcher, 1994). The host developmental program is not initiated in the absence of a symbiotic infection (McFall-Ngai and Ruby, 1991); however, once triggered, some (but not all) of the events will proceed without the continued presence of the bacteria (Doino and McFall-Ngai, 1995; Lamarcq and McFall-Ngai, 1998). As a result, a distinct, functional organ is formed through a program of biochemical accommodation and cooperation that is analogous to the well described development of the *Rhizobium*-legume symbiosis (van Rhijn and Vanderleyden, 1995).

Obstacles to Bacterial Colonization

To colonize an animal host, bacteria, whether pathogenic or cooperative, must overcome characteristics of the tissue that discourage the growth of inappropriate microorganisms. These obstacles to colonization may include general environmental conditions in the tissue, as well as active responses by the host's defenses. Five such challenges that appear to play a role in colonization of the light organ crypts are summarized below.

Viscous Barrier
Both flagellated and unflagellated motility mutants of *V. fischeri* are unable to initiate a colonization, even when presented as a 1000-fold higher inoculum, and for a 10-times longer exposure (Graf *et al.*, 1994). The cilia lining the duct leading to the crypts create an outward-directed current (McFall-Ngai and Ruby, 1998), and the presence of mucopolysaccharides in the ducts

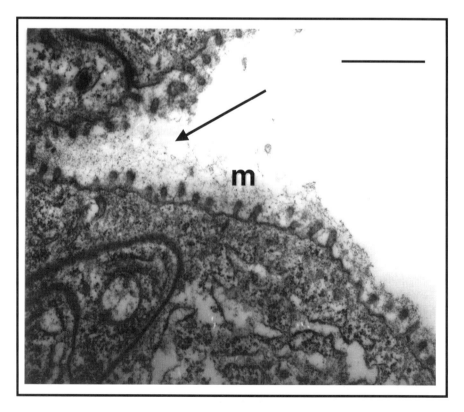

Figure 3. Mucus-like Material Around the Pore Leading to the Light Organ Crypts of *Euprymna scolopes*
Transmission electron micrographs reveal a flocculant, glycoprotein-like material (**m**) surrounding the ciliated cells that form the pore (arrow) leading into the crypt spaces (bar = 1 μm).

(Nyholm and McFall-Ngai, 1998) and emanating from the pore (Figure 3) suggests that the cells must penetrate a viscous barrier to reach the crypts themselves.

Nutritional Limitation

Proteins and/or peptides in the crypt contents may constitute a substantial source of nutrients: the ability of some amino acid auxotrophs to fully colonize the crypts indicates that the host must be a major source of these building blocks (Graf and Ruby, 1998). While it is not yet known whether there is a specific bacterial signal that is needed to elicit the host's production and export of these nutrients, two morphological observations of the colonized light organ are suggestive. First, the presence of goblet cells has been noted in both the duct and crypt regions of the light organ (Figure 2B). These goblet cells could be the source of mucopolysaccharide glycoproteins, like mucin, that might be secreted into the crypts. Second, the epithelial cells lining the crypt walls become edematous, swelling to about 4-fold the volume of the cells lining uninfected crypts (Montgomery and McFall-Ngai, 1994). Such swelling is indicative of osmotic stress in epithelial cells, which may lead to the leaking of organic materials into the surrounding extracellular spaces.

Clearing by Expulsion

The diurnal process of expulsion will select for symbionts that are able to attach to the epithelial cells lining the crypt spaces. Evidence exists to suggest that a mannose-sensitive adhesion is important for successful colonization (McFall-Ngai *et al.*, 1998; Feliciano and Ruby, 1999). In addition, the increasing intimacy observed between *V. fischeri* cells and the epithelial crypt microvilli as the association matures (Lamarcq and McFall-Ngai, 1998) suggests that bacterial outer membrane proteins (Aeckersberg and Ruby, 1998) or other surface components become increasingly important for persistent attachment.

Exposure to Reactive Oxygen Species

Recent data indicate that the uninfected crypt environment is oxidatively stressful; for instance, elevated superoxide anion production has been detected in the light organ tissue (Small and McFall-Ngai, 1993). In addition, the squid light organ has unusually high levels of mRNA encoding a HPO with significant sequence similarity to human myeloperoxidase, an antimicrobial protein in neutrophils (Tomarev *et al.*, 1993) that catalyzes the conversion of H_2O_2 and halide ions into microbicidal hypohalous acids. Activity assays and immunocytochemical analyses have demonstrated that HPO is present in the crypt space, and occurs at elevated levels elsewhere in the squid where there are pathogenic bacterial infections (Small and McFall-Ngai, 1998).

Removal by Phagocytes

One line of defense against colonization that *V. fischeri* cells must avoid appears to be the phagocytic cells in the crypts (Figure 2C) (Nyholm and McFall-Ngai, 1998). Such cells arise in a tissue called the white body and are typically found in the hemolymph (Budelmann *et al.*, 1997) and in extracellular spaces like the light organ crypts. Phagocytes of invertebrates share several ultrastructural, physiological and biochemical traits with mammalian cells such as neutrophils and macrophages that are associated with innate immunity (Sminia *et al.*, 1987; Greger *et al.*, 1996). These traits underlie the common responses of these cells to microorganisms, such as respiratory burst activity and phagocytosis. Such common response pathways, which can function through the release of chemokines and cytokines (*e.g.*, interleukins and tumor necrosis factor-alpha), are indicative of an ancient origin for the mechanisms by which animal cells and microorganisms interact (Belvin and Anderson, 1996; May and Ghosh, 1998). Continuing studies of these and other host-imposed barriers are an important focus of research into the basis of the reciprocal accommodation reached between *V. fischeri* and its host during the development of the light organ association.

Similarities between Benign and Pathogenic Tissue Colonization

In recent years it has become clearer that certain virulence determinants of pathogenic *Vibrio* species are expressed in, and may be required for, the benign *V. fischeri* colonization. The list of similarities between the colonization of host epithelium by *V. fischeri* and that of a number of pathogenic *Vibrio* species such as *V. vulnificus*, *V. parahaemolyticus* and *V. cholerae* continues to grow (Table 1). These similarities include known virulence determinants of *V. cholerae* (ToxR and CtxAB) that have protein homologs or analogs in *V. fischeri* (Reich and Schoolnik, 1994; Reich and Schoolnik, 1996; Reich *et al.*, 1997). Because of these similarities, and the phylogenetic relatedness of *V. fischeri* and pathogenic *Vibrio* species, it is reasonable to hypothesize that a number of properties of their associations with host tissue might have arisen from a common evolutionary origin. In fact, some species, like *V. vulnificus* and *V. harveyi*, can be a benign symbiont in one host and a virulent pathogen in another (DePaola *et al.*, 1994; Harris *et al.*, 1996) suggesting a duality of function in these different classes of association. This hypothesis is further supported by additional evidence of the homologous character of tissue colonization by *V. fischeri* and pathogenic *Vibrio* species. The identity of these *V. fischeri* genetic and enzymatic homologs suggests that they are important in three critical and interacting aspects of epithelial colonization: bacterial attachment, bacterial avoidance of host defense responses, and host provision of nutrition to the bacterial culture (Table 1). Because the commensal colonization of invertebrate tissues by *Vibrio* species arose millions of years before the development of human pathogenesis by members of this genus it seems likely that host-colonization mechanisms that were present in these cooperative associations may have been recruited into the development of pathogenic associations. A similar argument has been made to explain the existence of a majority of the known *Salmonella* virulence determinants in nonpathogenic *Escherichia coli*, the organism in which many of these determinants were first discovered and described (Groisman and Ochman, 1994)

The epithelial cell surfaces of the human body harbor a complex and essential community of bacterial species. These bacteria not only create a normal, healthy biochemical microenvironment on these surfaces, but also provide the first line of defense against invading pathogens. The growing awareness of the essential role that the natural microbiota play in the development and maintenance of the healthy state in animals (Hooper *et al.*, 1998) has led to an increasing recognition of the importance of benign bacterial colonization of tissues. To design effective therapeutic measures against potential diseases, we must understand the dynamic interactions between hosts and their essential microbial partners. However, little is known either about the signaling processes, and their genetic regulation, through which these beneficial bacteria establish stable associations with host cells, or about how these very same mechanisms are exploited by harmful bacteria.

The light organ symbiosis between *E. scolopes* and *V. fischeri* offers unique opportunities to study these processes. Important characteristics of this natural experimental model that make it particularly useful for comparisons of benign and pathogenic colonization of animal tissue are: (i) the laboratory culturability of both partners (Wei and Young, 1989; Boettcher and Ruby, 1990; Hanlon *et al.*, 1997); (ii) the specificity of the association (McFall-Ngai and Ruby, 1991); (iii) the similarity between the microvillous, polarized epithelial cells and phagocytes of the light organ crypts and those of the human intestinal lining (McFall-Ngai and Montgomery, 1990; Nyholm and McFall-Ngai, 1998); and, (iv) the close genetic relationship between *V. fischeri* and pathogenic relatives, such as *V. vulnificus* and *V. cholerae*.

As in most associations between bacteria and their hosts, this benign infection does not produce a diseased state in the colonized tissue; thus, the signals and responses of the two partners follow a predictable and integrated pattern. However, unlike the most common types of associations (*e.g.*, the normal microbiota of the enteric tract, skin and oral cavity), the *V. fischeri* light organ colonization is monospecific, thereby greatly simplifying the elucidation of bacteria-host interactions. To date, success with this model has resulted from collaborations involving bacterial molecular physiologists and geneticists, and animal developmental biochemists. By continuing to integrate efforts to understand the responses of each member of this association to the other, the chances of making important discoveries about bacterial tissue colonization will be enhanced.

Future Directions

Further investigations of the symbiotic colonization of *E. scolopes* by *V. fischeri* will promote a paradigm for studying host-bacterial interactions that will specifically serve to advance biomedical objectives in the following ways.
(i) Defining the biochemical and molecular events that can characterize the specific initiation, rapid colonization, and long-term persistence of benign bacterial infections is of increasing importance. While a number of other promising model systems are emerging (Forst and Nealson, 1996; Whittaker *et al.*, 1996; Hooper *et al.*, 1998; Graf, 1999), the squid-vibrio association is already becoming well-established (Ruby, 1996; McFall-Ngai, 1998b).

(ii) The importance of the kind of bacterial signaling called "quorum-sensing" (Fuqua *et al.*, 1996) in pathogenic bacterial-host signal interactions has become increasing evident. *V. fischeri*, the species in which this mechanism was first discovered, continues to be the paradigm system for the study of both intra- and interspecies association (Visick and Ruby, 1999).

(iii) Continued studies of bacterial and host determinants that potentiate light organ symbiosis may not only reveal further convergences with known pathogenic virulence factors (Table 1) but, more importantly, may also promote the discovery of as yet undescribed ones (Fidopiastis and Ruby, 1999). In addition, the level of understanding of the ecology of *V. fischeri* (Ruby and Lee, 1998) makes this species particularly well-positioned for use in testing experimental models of the evolution of horizontally transmitted pathogens (Ebert, 1998).

(iv) Future discoveries of the biology of the squid-vibrio association may aid in efforts to understand the mechanisms by which benign colonizations of mollusk tissue serve as a reservoir for human pathogenic *Vibrio* species (Tamplin and Capers, 1992), and to determine the role of such reservoirs in the recent resurgence of pathogenic epidemics (Colwell, 1996).

Acknowledgements

I thank M. McFall-Ngai and her laboratory for their critical collaborative interactions, and specifically for all the images used in this review. I also thank the members of my laboratory for producing much of the data I've discussed.

Portions of the research described herein were funded by the National Institutes of Health through grant RO1-RR12294 to E.G.R. and M. McFall-Ngai, and by National Science Foundation grant IBN96-01155 to M. McFall-Ngai and E.G.R.

References

Aeckersberg, F., and Ruby, E.G. 1998. Possible role of an outer membrane protein of *Vibrio fischeri* in its symbiotic infection of *Euprymna scolopes*. Abstr. Gen. Meet. Amer. Soc. Microbiol. 98: 374.

Belvin, M. P., and Anderson, K.V. 1996. A conserved signaling pathway: the *Drosophila* toll-dorsal pathway. Ann. Rev. Cell Dev. Biol. 12: 393-416.

Berry, S.S. 1912. The Cephalopoda of the Hawaiian Islands. Bull. U.S. Bur. Fish. 32: 255-362.

Blake, P.A. 1983. Vibrios on the half shell: what the walrus and the carpenter didn't know. Ann. Intern. Med. 99: 558-559.

Boettcher, K.J. 1994. Regulatory consequences of a high density bacterial population in the *Euprymna scolopes-Vibrio fischeri* light organ symbiosis. Doctoral dissertation, Univ. So. Calif. p. 1551.

Boettcher, K.J., and Ruby, E.G. 1990. Depressed light emission by symbiotic *Vibrio fischeri* of the sepiolid squid, *Euprymna scolopes*. J. Bacteriol. 172: 3701-3706.

Boettcher, K.J., and Ruby, E.G. 1995. Detection and quantification of *Vibrio fischeri* autoinducer from symbiotic squid light organs. J. Bacteriol. 177: 1053-1058.

Boettcher, K.J., Ruby, E.G., and McFall-Ngai, M.J. 1996. Bioluminescence in the symbiotic squid *Euprymna scolopes* is controlled by a daily biological rhythmn. J. Comp. Physiol. 179: 65-73.

Budelmann, B.U, Schipp, R., and Boletzky, S. v. 1997. Cephalopoda, Chp. 3. In: Microscopic Anatomy of the Invertebrates, Vol. 6A: Mollusca II. F. W. Harrison and A. J. Kohn, eds. Wiley, Liss, Inc., N.Y. p. 119-414.

Breznak, J.A., and Brune, A. 1994. Role of microorganisms in the digestion of lignocellulose by termites. Annu. Rev. Entomol. 39: 453-487.

Colwell, R.R. 1996 Global climate and infectious disease: the cholera paradigm. Science. 274: 2025-2031.

DePaola, A., Capers, G.M., and Alexander, D. 1994. Densities of *Vibrio vulnificus* in the intestines of fish from the U.S. Gulf Coast. Appl. Environ. Microbiol. 60: 984-988.

Doino, J.A., and McFall-Ngai, M.J. 1995. Transient exposure to competent bacteria initiates symbiosis-specific squid light organ morphogenesis. Biol. Bull. 189: 347-355.

Dougan, G. 1994. The molecular basis for the virulence of bacterial pathogens: implications for oral vaccine development. Microbiol. 140: 215-224.

Ebert, D. 1998. Experimental evolution of parasites. Science. 282: 1432-1435.

Feliciano, B., and Ruby, E.G. 1999. Identification and characterization of the *Vibrio fischeri* hemagglutination factor. Abstr. Gen. Meet. Amer. Soc. Microbiol. 99: 462.

Fidopiastis, P.M., Boletzky, S.v., and Ruby, E.G. 1998. A new niche for *Vibrio logei*, the predominant light organ symbiont of squids in the genus *Sepiola*. J. Bacteriol. 180: 59-64.

Fidopiastis, P.M., and Ruby, E.G. 1999. Cryptic luminescence in the cold-water fish pathogen *Vibrio salmonicida*. Arch. Microbiol. 171: 205-209.

Forman, H.J., and Thomas, M.J. 1986. Oxidant production and bacteriocidal activity of phagocytes. Ann. Rev. Physiol. 48: 669-680.

Forst, S., and Nealson, K.H. 1996. Molecular biology of the symbiotic-pathogenic bacteria *Xenorhabdus* spp. and *Photorhabdus* spp. Microbiol. Rev. 60: 21-43.

Foster, J. S., and McFall-Ngai, M.J. 1998. Induction of apoptosis by cooperative bacteria in the morphogenesis of host epithelial tissues. Dev. Genes Evol. 208: 295-303.

Fuqua, W.C., Winans, S.C., and Greenberg, E.P. 1996. Census and concensus in bacterial ecosystems: the LuxR-LuxI family of quorum-sensing transcriptional regulators. Annu. Rev. Microbiol. 50: 727-751.

Graf, J. 1999. Symbiosis of *Aeromonas veronii* biovar sobria and *Hirudo medicinalis*, the medicinal leech: a novel model for digestive tract associations. Infect. Immun. 67: 1-7.

Graf, J., Dunlap, P.V., and Ruby, E.G. 1994. Effect of transposon-induced motility mutations on colonization of the host light organ by *Vibrio fischeri*. J. Bacteriol. 176: 6986-6991.

Graf, J., and Ruby, E.G. 1994. The effect of iron-sequestration mutations on the colonization of *Euprymna scolopes* by symbiotic *Vibrio fischeri*. Abstr. Ann. Meet. Amer. Soc. Microbiol. 94: 76.

Graf, J., and Ruby, E.G. 1996. Characterization of a *Vibrio fischeri glnD* mutant: both iron and nitrogen utilization are affected. Abstr. Ann. Meet. Amer. Soc. Microbiol. 96: 509.

Graf, J., and Ruby, E.G. 1998. Host-derived amino acids support the proliferation of symbiotic bacteria. Proc. Natl. Acad. Sci. USA. 95: 1818-1822.

Greger, E., Drum, A., and Elston, R. 1996. Lucigenin- and luminol-dependent chemiluminescent measurement of oxyradical production in hemocytes of the Pacific razor clam, *Siliqua patula*, and the oyster, *Crassostrea gigas*, Chp. 16. In: Modulators of Immune Responses: The Evolutionary Trail. J.S. Stolen, T.C. Fletcher, C.J. Bayne, C.J. Secombes, J.T. Zelikoff, L.E. Twerkok, and D.P. Anderson, eds. SOS Publ., Fair Haven, N.J. p. 203-208.

Groisman, E.A., and Ochman, H. 1994. How to become a pathogen. Trends Microbiol. 2: 289-294.

Hanlon, R.T., Claes, M.F., Ashcraft, S.E., and Dunlap, P.V. 1997. Laboratory culture of the sepiolid squid *Euprymna scolopes*: a model system for bacteria-animal symbiosis. Biol. Bull. 192:364-374.

Harris, L., Owens, L., and Smith, S. 1996. A selective medium for *Vibrio harveyi*. Appl. Environ. Microbiol. 62: 3548-3550.

Hase, C.C., and Finkelstein, R.A. 1993. Bacterial extracellular zinc-containing metalloproteases. Microbiol. Rev. 57: 823-837.

Hooper, L.V., Bry, L., Falk, P.G., and Gordon, J.I. 1998. Host-microbial symbiosis in the mammalian intestine: exploring an internal ecosystem. BioEssays. 20: 336-343.

Jobling, M.G., and Holmes, R.K. 1997. Characterization of *hapR*, a positive regulator of the *Vibrio cholerae* HA/protease gene *hap*, and its identification as a functional homologue of the *Vibrio harveyi luxR* gene. Mol. Microbiol. 26: 1023-1034.

Jouravleva, E.A., McDonald, G.A., Marsh, J.W., Taylor, R.K., Boesman-Finkelstein, M., and Finkelstein, R.A. 1998. The *Vibrio cholerae* mannose-sensitive hemagglutinin is the receptor for a filamentous bacteriophage from *V. cholerae* O139. Infect. Immun. 66: 2535-2539.

Lamarcq, L.H., and McFall-Ngai, M.J. 1998. Induction of a gradual, reversible morphogenesis of its host's epithelial brush border by *Vibrio fischeri*. Infect. Immun. 66: 777-785.

Lee, K.-H., and Ruby, E.G. 1994a. Effect of the squid host on the abundance and distribution of symbiotic *Vibrio fischeri* in nature. Appl. Environ. Microbiol. 60: 1565-1571.

Lee, K.-H., and Ruby, E.G. 1994b. Competition between *Vibrio fischeri* strains during the initiation and maintenance of a light organ symbiosis. J. Bacteriol. 176: 1985-1991.

Lee, K.-H., and Ruby, E.G. 1995. Symbiotic role of the nonculturable, but viable, state of *Vibrio fischeri* in Hawaiian seawater. Appl. Environ. Microbiol. 61: 278-283.

May, M.J., and Ghosh, S. 1998. Signal transduction through NF-kappaB. Immunol. Today. 19: 80-88.

McFall-Ngai, M.J. 1990. Crypsis in the pelagic environment. Amer. Zool. 30:

175-188.

McFall-Ngai, M.J. 1994. Evolutionary morphology of a squid symbiosis. Amer. Zool. 34: 554-561.

McFall-Ngai, M.J. 1998a. The development of cooperative associations between animals and bacteria: establishing detente among Domains. Amer. Zool. 38: 593-608.

McFall-Ngai, M.J. 1998b. Pioneering the squid-vibrio model. ASM News. 64: 639-645.

McFall-Ngai, M.J., and Montgomery, M. 1990. The anatomy and morphology of the adult bacterial light organ of *Euprymna scolopes* Berry (Cephalopoda:Sepiolidae). Biol. Bull. 179: 332-339.

McFall-Ngai, M.J., and Ruby, E.G. 1991. Symbiont recognition and subsequent morphogenesis as early events in an animal-bacterial symbiosis. Science. 254: 1491-1494.

McFall-Ngai, M.J., and Ruby, E.G. 1998. Bobtail squid and their luminous bacteria: when first they meet. BioScience. 48: 257-265.

McFall-Ngai, M.J., Brennan, C., Weis, V., and Lamarcq, L. 1998. Mannose adhesin-glycan interactions in the *Euprymna scolopes-Vibrio fischeri* symbiosis. In: New Developments in Marine Biotechnology. Y. Le Gal and H. Halvorson, eds. Plenum Press, N.Y. p. 273-276.

Meighen, E.A. 1991. Molecular biology of bacterial bioluminescence. Microbiol. Rev. 55: 123-142.

Montgomery, M.K., and McFall-Ngai, M.J. 1992. The muscle-derived lens of a squid bioluminescent organ is biochemically convergent with the ocular lens. Evidence for recruitment of ALDH as a predominant structural protein. J. Biol. Chem. 267: 20999-21003.

Montgomery, M.K., and McFall-Ngai, M.J. 1993. Embryonic development of the light organ of the sepiolid squid *Euprymna scolopes*. Biol. Bull. 184: 296-308.

Montgomery, M.K., and McFall-Ngai, M.J. 1994. The effect of bacterial symbionts on early post-embryonic developmental of a squid light organ. Development. 120: 1719-1729.

Montgomery, M.K., and McFall-Ngai, M.J. 1998. Late postembryonic development of the symbiotic light organ of *Euprymna scolopes* (Cephalopoda:Sepiolidae). Biol. Bull. 195: 326-336.

Nealson, K.H., and Hastings J.W. 1991. The luminous bacteria. In: The Prokaryotes, a Handbook on the Biology of Bacteria: Ecophysiology, Isolation, Identification, Applications. 2nd ed. A. Balows, H.G. Truper, M. Dworkin, W. Harder, and K.H. Schleifer, eds. Springer-Verlag, N.Y. p. 625-639.

Nesis, K.N. 1987. Cephalopods of the World. T.F.H. Publications, Neptune City, N.J.

Nishiguchi, M., Ruby, E.G., and McFall-Ngai, M.J. 1998. Competitive dominance among strains of luminous bacteria provides an unusual form of evidence for parallel evolutionin sepiolid squid-vibrio symbioses. Appl. Environ. Microbiol. 64: 3209-3213.

Nyholm, S.V., and McFall-Ngai, M.J. 1998. Sampling the light organ microenvironment of *Euprymna scolopes*: Description of a population of host cells in association with the bacterial symbiont. Biol. Bull. 195: 89-97

Palmer, L.M., and Colwell, R.R. 1991. Detection of luciferase gene sequence in nonluminescent *Vibrio cholerae* by colony hybridization and polymerase chain reaction. Appl. Environ. Microbiol. 57: 1286-1293.

Reich, K.A., and Schoolnik, G.K. 1994. The light organ symbiont *Vibrio fischeri* possesses a homolog of the *Vibrio cholerae* transmembrane transcriptional activator ToxR. J. Bacteriol. 176: 3085-3088.

Reich, K.A., and Schoolnik, G.K. 1996. Halovibrin, secreted from the light organ symbiont *Vibrio fischeri*, is a member of a new class of ADP-ribosyltransferases. J. Bacteriol. 178: 209-215.

Reich, K.A., Beigel, T., and Schoolnik, G.K. 1997. The light organ symbiont *Vibrio fischeri* possesses two distinct secreted ADP-ribosyltransferases. J. Bacteriol. 179: 1591-1597.

Richardson, K. 1991. Roles of motility and flagellar structure in pathogenicity of *Vibrio cholerae*: analysis of motility mutants in three animal models. Infect. Immun. 59: 2727-2736.

Rijpkema, S.G.T., Bik, E.M., Jensen, W.H., Gielen, H., Versluis, L.F., Stouthamer, A.H., Guinee, P.A.M., and Mooi, F.R. 1992. Construction and analysis of a *Vibrio cholerae* δ-aminolevulinic acid auxotroph which confers protective immunity in a rabbit model. Infect. Immun. 60: 2188-2193.

Rook, G.A., and Stanford, J.L. 1998. Give us this day our daily germs. Immunol. Today. 19: 113-116.

Ruby, E.G. 1996. Lessons from a cooperative bacterial-animal association: The *Vibrio fischeri-Euprymna scolopes* light organ symbiosis. Annu. Rev. Microbiol. 50: 591-624.

Ruby, E.G. 1999. Ecology of a benign "infection": colonization of the squid luminous organ by *Vibrio fischeri*. In: Microbial Ecology and Infectious Disease. E. Rosenberg, ed. Amer. Soc. Microbiol. Press, Washington. p. 217-231.

Ruby, E.G., and Asato, L.M. 1993. Growth and flagellation of *Vibrio fischeri* during initiation of the sepiolid squid light organ symbiosis. Arch. Microbiol. 159: 160-167.

Ruby, E.G., and Lee, K.-H. 1998. The *Vibrio fischeri -Euprymna scolopes* light organ association: current ecological paradigms. Appl. Environ. Microbiol. 64: 805-812.

Ruby, E.G., and McFall-Ngai, M.J. 1992. A squid that glows in the night: development of an animal-bacterial mutualism. J. Bacteriol. 174: 4865-4870.

Salyers, A.A., and Whitt, D.D. 1994. Bacterial Pathogenesis: a Molecular Approach. Amer. Soc. Microbiol. Press, Washington, D.C.

Small, A.L., and McFall-Ngai, M.J. 1993. Changes in the oxygen environment of a symbiotic squid light organ in response to infection by its luminous bacterial symbionts. Amer. Zool. 33: 61A

Small, A.L., and McFall-Ngai, M.J. 1998. A halide peroxidase in tissues that interact with bacteria in the host squid *Euprymna scolopes*. J. Cellular Biochem. 72: 445-457.

Sminia, T., and van der Knaap, W.P.W. 1987. Cells and molecules in molluscan immunology. Dev. Comp. Immunol. 11: 17-28.

Sperandio, V., Giron, J.A., Silveira, W.D., and Kaper, J.B. 1995. The OmpU outer membrane protein, a potential adherence factor of *Vibrio cholerae*.

Infect. Immun. 63: 4433-4438.

Stabb, E.V., and Ruby, E.G. 1998. Construction of a mobilizable Tn5 derivative and its utility for mutagenizing, cloning and mapping loci in the luminescent symbiont *Vibrio fischeri*. Abstr. Ann Meet. Amer. Soc. Microbiol. 98: 375.

Tamplin, M.L., and Capers, G.M. 1992. Persistence of *Vibrio vulnificus* in tissues of Gulf Coast oysters, *Crassostrea virginica*, exposed to seawater disinfected with UV light. Appl. Environ. Microbiol. 58: 1506-1510.

Tashima, K.T., Carroll, P.A., Rogers, M.B., and Calderwood, S.B. 1996. Relative importance of three iron-regulated outer membrane proteins for *in vivo* growth of *Vibrio cholerae*. Infect. Immun. 64: 1756-1761.

Tomarev, S.I., Zinovieva, R.D., Weis, V.M., Chepelinsky, A.B., Piatigorsky, J., and McFall-Ngai, M.J. 1993. Abundant mRNAs in the bacterial light organ of a squid encode a protein with high similarity to mammalian antimicrobial peroxidases: Implications for mutualistic symbioses. Gene. 132: 219-226.

van Rhijn, P., and Vanderleyden, J. 1995. The *Rhizobium*-plant symbiosis. Microbiol. Rev. 59: 124-142.

Visick, K.L., Orlando, V.I., and Ruby, E.G. 1996. Role of the *luxR* and *luxI* genes in the *Vibrio fischeri-Euprymna scolopes* symbiosis. Abstr. Ann. Meet. Amer. Soc. Microbiol. 96: 509.

Visick, K.L., and Ruby, E.G. 1996. Construction and symbiotic competence of a *luxA*-deletion mutant of *Vibrio fischeri*. Gene. 175: 89-94.

Visick, K.L., and Ruby, E.G. 1997. New genetic tools for use in the marine bioluminescent bacterium *Vibrio fischeri*. In: Bioluminescence and Chemiluminescence, Proceedings of the 9th International Symposium 1996. J.W. Hastings, L.J. Kricka and P.E. Stanley, eds. Wiley and Sons, N.Y. p. 119-122.

Visick, K.L., and Ruby, E.G. 1998a. Temporal control of *lux* gene expression in the symbiosis between *Vibrio fischeri* and its squid host. In: New Developments in Marine Biotechnology. Y. Le Gal and H. O. Halvorson, eds. Plenum Press, N.Y. p. 277-279.

Visick, K.L., and Ruby, E.G. 1998b. The periplasmic, group III catalase of *Vibrio fischeri* is required for normal symbiotic competence and is induced both by oxidative stress and approach to stationary phase. J. Bacteriol. 180: 2087-2092.

Visick, K.L., and Ruby, E.G. 1999. The emergent properties of quorum sensing: consequences to bacteria of autoinducer signaling in their natural environment. In: Cell-Cell Signaling in Bacteria. G.M. Dunny and S.C. Winans, eds. Amer. Soc. Microbiol. Press, Washington, D.C. p. 333-352.

Wei, S.L., and Young, R.E. 1989. Development of symbiotic bacterial luminescence in a nearshore cephalopod, *Euprymna scolopes*. Mar. Biol. 103: 541-546.

Weis, V.M., Small, A.L., and McFall-Ngai, M.J. 1996. A peroxidase related to the mammalian antimicrobial protein myeloperoxidase in the *Euprymna-Vibrio* mutualism. Proc. Natl. Acad. Sci. USA. 93: 13683-13688.

Weis, V., Montgomery, M.K., and McFall-Ngai, M.J. 1993. Enhanced production of ALDH-like protein in the bacterial light organ of the sepiolid squid *Euprymna scolopes*. Biol. Bull. 184: 309-321.

Whittaker, C.J., Klier, C.M., and Kolenbrander, P.E. 1996. Mechanisms of adhesion by oral bacteria. Annu. Rev. Microbiol. 50: 513-552.

Wu, Z., Milton, D., Nybom, P., Sjo, A., and Magnusson, K.-E. 1996. *Vibrio cholerae* hemaglutinin/protease (HA/protease) causes morphological changes in cultured epithelial cells and perturbs their paracellular barrier function. Microbial Path. 21: 111-123.

4

Bacterial Signals and Antagonists: The Interaction Between Bacteria and Higher Organisms

Scott A. Rice[1,2], Michael Givskov[3],
Peter Steinberg[2,4], and Staffan Kjelleberg[1,2]

[1]The School of Microbiology and Immunology, The University of New South Wales, Sydney NSW 2052, Australia. [2]The Centre for Marine Biofouling and Bio-Innovation, The University of New South Wales, Sydney NSW 2052, Australia. [3]The Department of Microbiology, The Technical University of Denmark, DK-2800 Lynby, Denmark. [4]The School of Biological Sciences, The University of New South Wales, Sydney NSW 2052, Australia

Abstract

It is now well established that bacteria communicate through the secretion and uptake of small diffusable molecules. These chemical cues, or signals, are often used by bacteria to coordinate phenotypic expression and this mechanism of regulation presumably provides them with a competitive advantage in their natural environment. Examples of coordinated behaviors of marine bacteria which are regulated by signals include swarming and exoprotease production, which are important for niche colonisation or nutrient acquisition (*e.g.* protease breakdown of substrate). While the current focus on bacterial signalling centers on N-Acylated homoserine lactones, the quorum sensing signals of some Gram-negative bacteria, these are not the only types of signals used by bacteria. Indeed, there appears to be many other types of signals produced by bacteria and it also appears that a bacterium may use multiple classes of signals for phenotypic regulation. Recent work in the area of marine microbial ecology has led to the observation that some marine eukaryotes secrete their own signals which compete with the bacterial signals and thus inhibit the expression of bacterial signalling phenotypes. This type of molecular mimicry has been well characterised for the interaction of marine prokaryotes with the red alga, *Delisea pulchra*.

Introduction

Bacteria have developed numerous adaptations which help them to sense and respond to their environment; some classic stimuli which have been extensively studied include temperature shifts, nutrient availability, hydrostatic pressure, oxygen, and pH. Importantly, it is this ability to respond or interact with the local environment which determines the overall success or failure of an organism. Microorganisms are also capable of sensing and responding to the presence of other organisms. Clearly it is essential for an organism to sense the levels of its own population as well as other members of the community so that the organism can take the appropriate action. One mechanism by which bacteria sense and respond to members of the community is through the secretion and uptake of small diffusable chemical cues or signalling molecules. This includes density dependent population monitoring, in which the organism takes a population census, determined by the concentration of signalling molecules present, and when the required cell density, or quorum, is detected, the phenotype is expressed (Fuqua *et al.*, 1996; Fuqua *et al.*, 1994; Swift *et al.*, 1996).

It is becoming increasingly clear that many different microorganisms rely on signal sensing mechanisms, including quorum sensing, to regulate a variety of adaptive phenotypes. Signalling molecules come in a wide range of classes and types from a large variety of bacteria including but not restricted to: amino acids, *Myxococcus xanthus* (Kuspa *et al.*, 1992a; Kuspa *et al.*, 1992b); cAMP, *Dictyostellium discoidium* (Van Haastert, 1991), short peptides, *Enterococcus faecalis* (Clewell and Weaver, 1993); cyclic dipeptides (CDPs), *Pseudomonas aeruginosa* (Holden *et al.*, 1999); butanolides, *Streptomyces* (Yamada *et al.*, 1987); fatty acid derivatives, *Ralstonia solonacerum* (Flavier *et al.*, 1997); acylated homoserine lactones (AHLs) *e.g. V. fischeri*, *V. anguillarum* and *Aeromonas hydrophila* (Eberhard *et al.*, 1981; Milton *et al.*, 1997; Swift *et al.*, 1997), and there are some organisms with as yet uncharcterised signalling molecules, *e.g. V. harveyi*, *V. vulnificus*, and *V. angustum* (Bassler *et al.*, 1994; deNys *et al.*, 1999). Moreover, it should be pointed out that bacteria are not limited to one single signalling type. In the case of *V. harveyi* (see below), there are two signalling systems which interact to regulate bioluminescence and *R. solonacerum* uses 3-hydroxy palmitic acid methyl ester as a density dependent signal to regulate exopolysaccharide production through the regulation of at least three two component regulatory systems (Huang *et al.*, 1998).

As expected, just as the list of signalling systems is wide and varied, so too are the phenotypes which are regulated by those signalling systems. Examples of those phenotypes include: sporulation (butanolides), fruiting body formation (amino acids, cAMP), development of competence (peptides), starvation adaptation (generally uncharacterised), bioluminescence (AHLs), pigment production (AHLs), colonisation phenotypes (AHLs), virulence factor production (AHLs and an unknown factor in *Salmonella typhimurium*), and for some signals, the associated phenotypes are poorly described (CDPs). While most of the sensing systems described above seem to be associated with a single organism or phenotype, two systems, the AHL and CDP systems, appear to be broadly disseminated in a vast number of bacteria. A study by

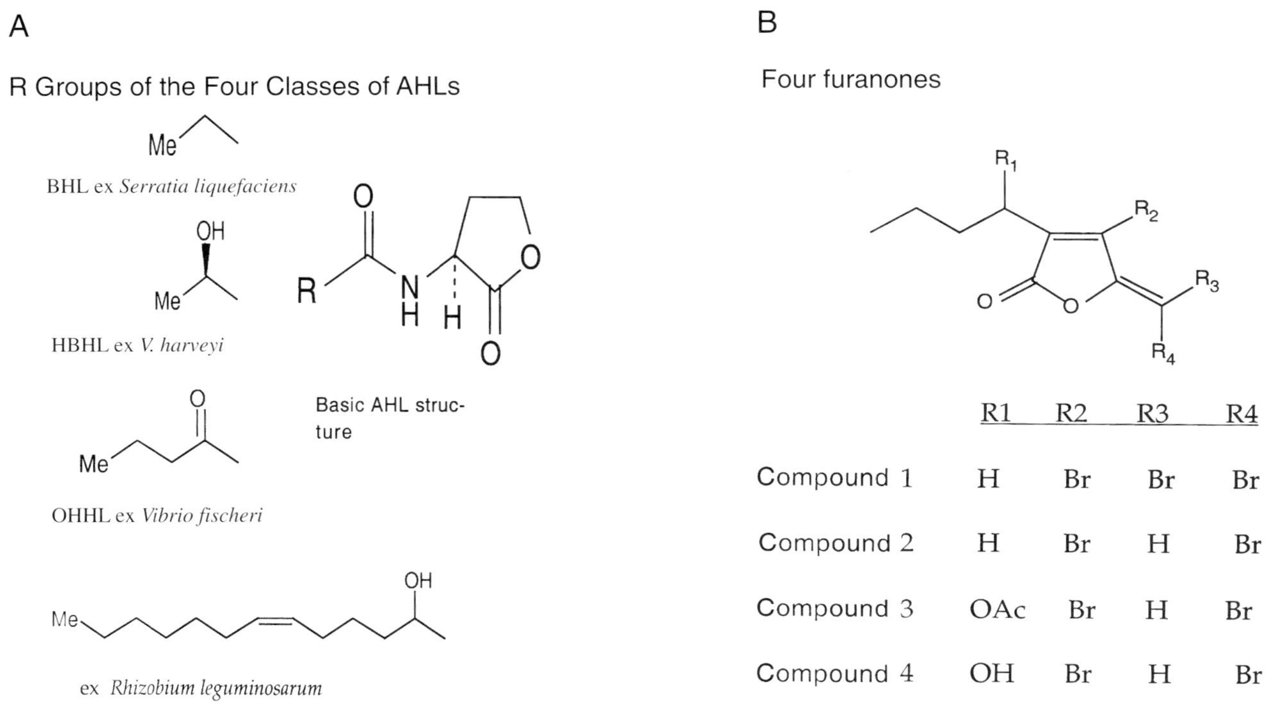

A

R Groups of the Four Classes of AHLs

BHL, ex *Serratia liquefaciens*

Me⌐⌐

HBHL, ex *V. harveyi*

Me⌐⌐OH

OHHL, ex *Vibrio fischeri*

Me⌐⌐⌐O

ex *Rhizobium leguminosarum*

Me⌐⌐⌐⌐⌐OH

R⌐C(=O)⌐N(H)⌐[lactone ring with C=O and O]

Basic AHL structure

B

Four furanones

[furanone structure with R1, R2, R3, R4 substituents]

	R1	R2	R3	R4
Compound 1	H	Br	Br	Br
Compound 2	H	Br	H	Br
Compound 3	OAc	Br	H	Br
Compound 4	OH	Br	H	Br

Figure 1. The Structures and Molecular Comparison of Bacterial Signals (AHLs) and Eukaryotic Signal Antagonists (Furanones) A. AHLs can be grouped into four classes based on the substitutions to the acyl chain. Group 1, represented by BHL, have no substitutions. Group 2, represented by HBHL, have a hydroxy group at the 3' position. Group 3, represented by OHHL, have an ester linked oxygen at the 3' position. Group 4 have an unsaturated bond in the acyl chain. B. The basic furanone structure is presented with R groups which represent different substitutions which are indicative of the members of the chemical family.

Cha *et al.* (1998) demonstrated that approximately 60% of all soil isolates, including *Agrobacteria, Rhizobia, Pseudomonas, Pantoea,* and *Erwinia* species, produce extracellular molecules which activate AHL bioassays. Similarly, Gram *et al.,* (1999) have demonstrated that 75% of *Enterobacteriaceae* isolated from processed sea-food produce AHLs. Marine isolates are no exception, as studies have demonstrated that approximately 10% of marine bacteria from the surface of the red alga *Delisea pulchra* produce AHLs (Maximilien *et al.,* 1998; England *et al.* unpublished). We have identified CDPs in supernatants from *P. aeruginosa, Proteus mirabilis,* and *Citrobacter freundii* (Holden *et al.* 1999). Furthermore, it has been previously reported that numerous organisms produce CDPs (Prasad, 1995), and in particular, many marine bacteria have been associated with CDP production, including a *Micrococcus sp.* (Stierle *et al.,* 1988) and *Vibrio parahaemolyticus* (Bell *et al.,* 1994).

AHL Mediated Sensing

Besides AHLs being widely prevalent among Gram negative bacteria, the AHL family of quorum sensing systems share some highly conserved features. Some details of AHL sensing systems are included here as background to the subsequent discussion of cross-talk of signals produced from other sources, with a specific focus on signals from the marine environment. AHLs consist of a lactone ring covalently linked to an acylated fatty acid chain through an amide bond (Figure 1A). The various AHLs produced differ in the length and the degree of saturation of the acyl chain. *Vibrio fischeri* serves as the paradigm for AHL mediated quorum sensing. Central to the AHL family of quorum sensing genes are a pair of genes, *luxI* and *luxR* and their homologues, which respectively code for the AHL synthase, LuxI, and the membrane bound AHL receptor, LuxR, which also acts as a transcription factor. In essence, *luxI* is transcribed at a low-basal rate, which allows for a constant low rate of production of the AHL signal. As the population density increases, the AHL concentration increases until a critical threshold concentration is reached, usually during late log or early stationary phase, at which time, there is sufficient AHL present to bind to LuxR. The AHL-LuxR complex is then capable of stimulating transcription of *luxI* which in turn leads to even more AHL production, thus forming a positive feedback or autoinduction circuit. This autoinduction circuit is the key to quorum sensing as it allows for a rapid increase in signal production and dissemination and therefore leads to a rapid induction of the phenotype by all members of the population. In addition to stimulating transcription of the *luxI*, the AHL-LuxR complex also stimulates transcription of the quorum sensing phenotypic genes.

   The interaction of the AHL with the cognate receptor shows a range of specificity and current data indicate that there is a structure-function relationship for cross reaction of AHLs with the receptors. For *V. fischeri* the interaction is very specific, the LuxR protein optimally responds to the cognate AHL, OHHL. In contrast, the LuxR homologues in *Chromobacterium violaceum* and *Agrobacterium tumefaceins* can bind to a number of different AHLs in addition to their own native signals. This property also makes these strains useful as bioassays for the detection of signals produced by other organisms (McClean et al., 1997a; 1997b; Shaw et al., 1997).

However, there are exceptions to all models, and AHL quorum sensing is no exception. Bioluminescence is also regulated by a quorum sensing mechanism in a closely related bacterium, V. harveyi, although the mechanism of quorum sensing is quite different. V. harveyi has two signalling systems, one of which produces and responds to the AHL, HBHL, while the second system responds to an as yet unidentified chemical cue. While the exact identity of the non-AHL signal molecule remains unknown, recent work by Surette et al. (1999) has identified that a mutation in the luxS gene of V. harveyi abolishes production of the non-AHL signal in that bacterium, suggesting that luxS is the gene required for synthesis of the signal. Genome database searches have also determined that the luxS genes are widespread among a variety of Gram-negative and even some Gram-positive bacteria (Surette et al., 1999) which would suggest that the non-AHL signaling system is widespread amongst bacteria. In addition, the response regulator, LuxN, that binds HBHL is not a homologue of the V. fischeri LuxR, but rather is the membrane bound component of a two component signal transduction pathway (Bassler and Silverman, 1995). The second signalling system appears to also be sensed by another two component regulatory pair, LuxQ and LuxP (Bassler and Silverman, 1995). Signalling information from the two pathways is integrated through a common transcriptional regulator, LuxO. Phosphorylation of LuxO, allows for derepression of the bioluminescence genes, which are then transcriptionally activated by LuxR; it should be noted that the LuxR of V. harveyi shares no sequence homology to the LuxR of V. fischeri (Bassler and Silverman, 1995).

Regulation of AHL phenotypes is primarily controlled by binding of the LuxR-AHL complex to the promoter region of the genes involved, however, this is not the only level of regulation for either the phenotypic genes or the luxIR homologues. There is a growing body of evidence indicating that many global regulators are important in the control of quorum sensing genes. Catabolite repression (Ulitzur and Dunlap, 1995) has been implicated in the regulation of lux genes of V. fischeri as has the histone-like protein, HNS (Ulitzur et al., 1997) and the heat shock sigma factor, σ^{32} (Ulitzer and Kuhn, 1988). In Pseudomonas spp. regulation of the las system is controlled by a host of global regulators including: vfr, a catabolite repressor protein homologue (Albus et al., 1997) and a two component regulatory pair, gacA and lemA (gacS) (Kitten et al., 1998). In addition, the second quorum sensing system in P. aeruginosa, the rhl system, which is partially regulated by the las system, regulates the expression of the global and stationary phase sigma factor, rpoS (Latifi et al., 1995; Latifi et al., 1996). Thus, it seems that signalling systems, like many bacterial responses, are regulated through multiple pathways and in this way can fine tune their expression of the relevant phenotypes.

Implicit in the model is the ability of the AHLs to rapidly diffuse into or out of the cell across the membrane. It was previously demonstrated in V. fischeri that AHLs can freely diffuse across the bacterial membrane (Kaplan and Greenberg, 1985). Therefore, if diffusion of AHLs out of the cells is not impeded and AHL synthesis occurs at a particular constant rate prior to autoinduction, then quorum regulation is directly correlated to the cell density of the population or more correctly, in a given environment (*e.g.* a culture flask), the threshold concentration of AHL required for quorum sensing will be achieved at a reproducible cell density. However, the suggestion has been put forth that not all AHLs diffuse rapidly across the membrane. It is conceivable that some of the AHLs such as OdDHL which have longer acyl chains are less likely to diffuse across the membrane and therefore may require a transport mechanism to cross the cell membrane. To date, there have been no demonstrations of an AHL transport system, although Evans *et al.* (1998) demonstrated that overexpression of a MexAB-OprM multi-drug efflux pump in *P. aeruginosa* will transport autoinducer out of the cell and repress AHL mediated phenotypes. However, as this is over expression of a non-specific transporter, it is doubtful that it normally plays a role in transport of AHLs. If indeed there is some diffusion barrier which slows or limits export of AHLs, then it may be possible to induce the density dependent phenotype at low cell densities or independent of cell density. One possible environment where low cell density induction might occur is in a population of cells growing as a biofilm where there may be differences in autoinducer accumulation in the biofilm or diffusion of the AHL out of the cell into the biofilm. Similarly, changes in the cell membrane which effect its fluidity, as in response to stress conditions such as osmotic or heat shock or starvation may also decrease the rate of diffusion of AHLs out of the cell and cause induction at low cell density.

Using mathematical models, we have predicted a number of factors which might lead to early autoinduction at low cell densities (Nilsson *et al.*, unpublished). Some of these factors include low growth rates, a very high rate of autoinduction, and the biofilm acting as a diffusion barrier between the cell membrane and the biofilm. In some scenarios, such as low diffusion rates and slow growth, we predict that maximum autoinduction could occur at low cell densities when cells are in the early logarithmic phase of growth. While there is evidence that signals are produced in biofilm settings (McClean *et al.*, 1997b; Stickler *et al.*, 1998), there are few studies that have directly measured the concentration of AHLs or population density required to induce a phenotype in its natural setting and likewise, there are few studies that measure the induction of AHL mediated quorum sensing genes in biofilms. These concepts of cell-cell signalling, diffusion of signals, and activation at low cell densities in restricted environments are particularly relevant for marine bacteria, especially in the context of adhesion and colonisation phenotypes, where a single cell may interact with a surface.

Since bacteria use diffusable signalling molecules to regulate phenotypes in the environment, and these responses often involve a eukaryotic host, then it does not seem unlikely that eukaryotic organisms would have developed mechanisms to exploit those bacterial signals for their own benefit. There is evidence for such interactions between the light organs of squid and fish to attract and control their bioluminescent symbiont (see the review by Ruby in this issue) and there is evidence that plants may produce extra-cellular factors that induce AHL mediated phenotypes in *Agrobacterium* (Piper *et al.*, 1993; Zhang *et al.*, 1993). While the prokaryote-eukaryote interaction in these examples is beneficial to both, there are clearly instances where the prokaryote is damaging to the eukaryotic host. For example, there is evidence that OdDHL, produced by *P. aeruginosa*, can shift the host immune response from a Th1, cell mediated antibacterial response, to a Th2 response, which is less effective at clearing bacterial pathogens (Telford *et al.*, 1998). In this way, it is supposed that the pathogen modifies the host environment so that it might be better able to compete and survive. Therefore it would be advantageous to the eukaryote host to be able to inhibit bacterial signalling phenotypes which include the expression of virulence factors as well as colonisation traits.

Signals from *Delisea pulchra* Cross-talk with AHL Systems

An elegant example of eukaryotic interference with bacterial signalling processes comes from the control of bacterial fouling on the marine macro-algae, *Delisea pulchra*. This red alga originally came to our attention because it did not become fouled by other marine invertebrates or plants even though other plants in the same area were heavily fouled. Fouling is damaging to marine plants in that it decreases their photosynthetic ability, increases drag in flowing water, which leads to tearing of the plant, and fouling organisms often parasitise the plant. Bacteria are important in the initial colonisation of surfaces by macroorganisms in that they provide a conditioning film which attracts other organisms to the site. When the surface of *D. pulchra* was examined more closely, it was observed that just as there were fewer fouling macroorganisms, there were also fewer bacteria on the surface of the plant with the tips of the plant having the fewest number of bacteria and the base or stem of the plant having the highest numbers (Maximilien *et al.*, 1998). Investigations into the strategy by which *D. pulchra* remains free of fouling organisms led to the discovery that *D. pulchra* produces a number of halogenated furanone compounds which are secreted onto the surface of the plant at concentrations ranging from 1-100 ng/cm^2 (deNys *et al.*, 1998; deNys *et al.*, 1993; Kazlauskas *et al.*, 1977). The variation of bacterial numbers inversely correlates with the surface concentration of furanones on the different parts of the plant with the tips having the highest furanone concentration and the lowest bacterial concentration (Maximilien *et al.*, 1998). Structurally, furanones are similar to AHLs which led to the hypothesis that furanones might specifically interfere with AHL mediated phenotypes. The alga primarily produces 4 furanones and approximately 10 others in minor quantities (Figure 1B). These halogenated furanones consist of a five carbon furan ring structure with an exocyclic double bond at the C-5 position and a substituted acyl chain at the C-3 position (Figure 1B) (deNys *et al.*, 1993). Molecular modeling of furanones and AHLs demonstrated the similarity of furanones and AHLs by comparing stick models of the most stable configurations for the molecules (not shown). The ring structures and the acylated side chain match quite closely and the substituted side-chains align and extend in similar fashion. As it is the acylated side chain of the AHL that gives it its specificity of interaction with a LuxR homologue, we anticipate that, if furanones interfere with AHL phenotypes, that furanones will show differential activity against AHL mediated phenotypes and that such differences will be reflected in the substitutions of the R groups.

When plastic coupons coated with furanones are incubated in the marine environment and compared with non-furanone controls, a concentration dependent inhibition of bacterial attachment to those coupons was observed (Kjelleberg *et al.*, 1997). Importantly, the furanone concentrations that inhibited attachment were similar to those presented on the surface of the plant in the natural environment. Similarly, when a number of marine isolates were assayed for their ability to swarm in the presence of furanones, swarming was inhibited in a concentration dependent manner. This data suggested that furanones were produced by the plant and secreted onto its surface to specifically inhibit the bacterial colonisation phenotypes of attachment and swarming. Importantly, swarming is a high cell density phenotype that is known to be regulated by AHLs (Eberl *et al.* 1996) and attachment and biofilm formation has recently been suggested to be controlled by AHL systems (Davies *et al.*, 1998). Therefore, based on the structural similarities of furanones with AHLs and the effect of furanones on controlling the AHL mediated colonisation phenotypes of attachment and swarming, it was hypothesised that furanones cause these effects by specifically inhibiting AHL mediated gene regulation.

To approach this hypothesis, the effect of furanones in a number of AHL bioassays was tested. Furanone mediated inhibition of bioluminescence was monitored using either wild-type strains of *V. fischeri* and *V. harveyi* or by using a plasmid based reporter system (pSB403) in *E. coli* (Winson *et al.*, 1998). The plasmid reporter pSB403 lacks an AHL synthase gene, thus AHLs need to be added exogenously to induce bioluminescence. For all three systems, it was demonstrated that furanones dramatically inhibited bioluminescence (Figure 2A). Similarly, furanones inhibited swarming in *Serratia liquefaciens* at low concentrations and the degree of swarming inhibition was concentration dependent (Givskov *et al.*, 1996; Manefield *et al.*, 1999*).* Specifically, furanone compound 2 was demonstrated to reduce the swarming velocity of *S. liquefaceins* by >70% at 10 μg/ml and completely inhibited swarming at 100 μg/ml (Givskov *et al.* 1996). The con-centrations used to inhibit bioluminescence and swarming in these assays were well within the range of concentrations presented at the surface of the plant (100 μg/ml which is approximately equal to 100 ng/cm^2) (deNys *et al.*, 1998). Swarming of *S. liquefaceins* is particularly relevant in the marine setting as a number of bacteria isolated from the surface of *D. pulchra* have been demonstrated to exhibit a swarming phenotype and furanones were effective at inhibiting swarming of those marine isolates (Maximilien *et al.*, 1998). For both the swarming and the bioluminescence, different furanones had differential effects, with some furanones being strongly active and some having little or no effect (Givskov *et al.* 1996).

Some bacteria rely on proteases for the invasion and colonisation of a host; it has also been demonstrated that in *S. liquefaceins* the activity of two exoproteases is regulated in a density dependent fashion (Givskov *et al.* unpublished). Addition of furanones to cultures of *S. liquefaceins* also decreases the activity of these two exoproteases. The prawn pathogen *Vibrio harveyi*, which has two quorum sensing systems, regulates its toxicity in a density dependent manner and we have demonstrated that extra-cellular supernatants show reduced toxicity in prawns when *V. harveyi* was

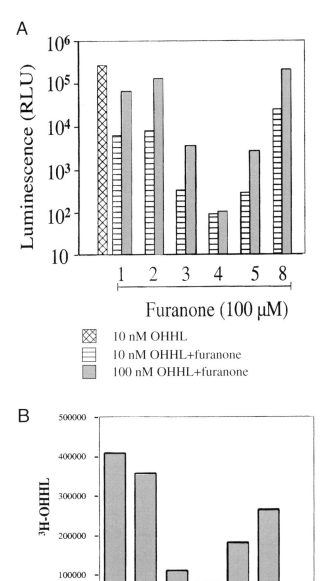

Figure 2. The Effect of Furanones in AHL Bioassays
A. The effect of furanones on the bioluminescence of an *E. coli* strain carrying the bioluminescent monitor plasmid pSB403. Inhibition of bioluminescence in the presence of furanones is normalised against level of bioluminescence in the presence of 10 nM OHHL (100%). Two concentrations of OHHLs were used, 10 nM and 100 nM, while the furanone concentration was held constant at 100 µM. B. Addition of furanones to a LuxR overproducing system displaces binding of the native signal, OHHL, in a concentration dependent manner. The degree of displacement is monitored by the amount of ^{3}H-OHHL that binds to the LuxR producing strain in the presence of furanones. Note that the degree of displacement mirrors the degree of bioluminescence inhibition in that compound 4 is the most effective at displacement and is the most effective at inhibiting both bioluminescence.

preincubated in the presence of furanones. This reduction in toxicity correlated to the reduction of two exoproducts as determined by PAGE analysis (Harris *et al.* unpublished). In all of the examples described here, the furanones were able to inhibit the AHL phenotypes in the presence of the native inducer, either produced naturally in the wild-type system or added exogenously to strains carrying the monitor plasmid or to genetically engineered strains which lack a functional AHL synthase gene (*luxI* homologue). This data would suggest that the furanones competitively inhibit AHLs from binding the LuxR receptor and thus prevent expression of the signalling phenotype.

To specifically address the mode of action, the effect of furanones on the global metabolism of the cells was assessed. At the concentrations tested, none of the furanones used were growth inhibitory to the bacteria in the bioassays. However, it was still possible that furanones effected the energy levels of the cells which might explain a decrease in bioluminescence, which is an energy intensive process. To test this, cells harboring a monitor plasmid which constitutively expresses the *lux* genes were incubated in the presence of furanones (Kjelleberg *et al.*, 1997). If furanones cause gross aberrations in energy metabolism, there should be a decrease in luminescence when these cells are grown in the presence of furanones, which was not observed. This observation would indicate that the furanones do not cause a significant reduction in the cellular pool of ATP. The possibility still remained that inhibition is through some other non-specific mechanism such as protein regulation; for example, furanones could globally inhibit transcription or translation. Therefore, the effect of furanones on protein production was examined by looking at total protein production by 2 dimensional PAGE analysis. *E. coli* cells harboring *lux* monitor plasmids were incubated in the presence or absence of furanones as well as in the presence or absence of the autoinducer OHHL, the proteins were extracted and the protein patterns were compared for differences. Addition of OHHL resulted in an induction of three proteins which correspond to the pSB403 encoded proteins LuxA, LuxB, and LuxD (Manefield *et al.*, 1999). Incubation of the *E. coli* monitor strain in the presence of furanones resulted in the inhibition of six proteins, three of which corresponded to LuxA, LuxB, and LuxD, and six proteins were induced. Comparison with the *E. coli* proteome data base suggested that the other three repressed proteins were OmpF, DnaK, and glutamate ammonia ligase. Of the six induced, four are unidentified, while the remaining two were glucose-6-phosphate dehydrogenase, and alkylhydroperoxide reductase. This again indicated that furanones were not causing gross aberrations in cellular metabolism, but rather were having very specific effects on the AHL regulatory pathway. This hypothesis was supported by the molecular modeling data which shows that furanones and AHLs are very similar molecules. To rule out furanone mediated modification of the AHL, proton NMR spectroscopy of mixtures of furanones and AHLs was performed and no modification of either compound observed, which confirmed that there was no modification of AHLs by the furanones (Manefield *et al.*, 1999).

If the furanones do not effect global metabolism and do not alter the physical-chemical properties of the autoinducer, and if the furanones are structurally similar to AHLs then it is probable that they mediate their effects by competing for the same regulator, LuxR. This theory is supported by

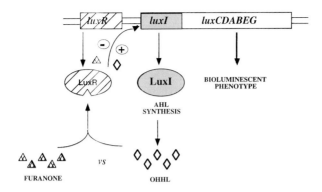

Figure 3. A Model for the Mechanism of Furanone Mediated Inhibition of AHL Phenotypes
In the absence of furanones, AHLs, produced by the LuxI protein, bind the receptor, LuxR, and induce transcription, noted in the figure as a "+", of *luxI* and the genes involved in development of the quorum sensing phenotype (e.g. bioluminescence). Based on the structural similarity of furanones and AHLs, their ability to inhibit AHL phenotypes, and their ability to displace the native AHL signal, we suggest that furanones competitively bind the LuxR receptor and prevent transcription of the phenotypic genes, noted in the figure as a "-".

displacement experiments in which the binding of ^{3}H labeled OHHL to LuxR overproducing cells is reduced in a concentration dependent fashion by the presence of furanones (Figure 2B) (Manefield *et al.*, 1999). Similarly, the effects of the furanones could be overcome by adding back the OHHL in excess. The strength of displacement mirrored the degree of inhibition of bioluminescence by the furanone, indicating a correlation between furanone binding and inhibition of AHL phenotypes. Thus, it would appear that furanone mediated inhibition of AHL phenotypes occurs as a direct result of competition by the furanone for the AHL binding site on the LuxR receptor-transcriptional activator and leads to the model for inhibition described in Figure 3. In this model, furanones and AHLs compete for binding to the receptor site of the transcriptional activator, LuxR. When AHLs bind the receptor, the transcriptional activator can induce expression of LuxI and the phenotypic genes. In contrast, with furanones bound, LuxR is not able to activate transcription and the system remains in the off position.

Therefore, based on the available data, it appears that the marine red alga *D. pulchra* has evolved a mechanism whereby it can exploit quorum sensing regulation of bacteria to reduce the levels of bacteria on its surface which helps to alleviate problems associated with fouling. The alga does so through molecular mimicry by producing signal analogues which competitively interfere with the bacterium's endogenous signal to bind its cognate receptor. This also provides the first example of inhibitory signal cross-talk in the marine environment. However, cross-talk is not likely to be limited solely to furanones, indeed, our laboratory has isolated and structurally characterised several new classes of molecules from marine algae which are also capable of cross-talking with known bacterial signalling phenotypes (deNys *et al.* unpublished).

Other Novel Signals and Cross-talk

To identify novel bacterial signal systems which cross-talk with AHL quorum sensing systems, we have screened supernatants from a number of marine and non-marine bacteria using AHL based bioassays. One class of signalling molecules we have recently identified, diketo-piperazines or cyclic dipeptides (CDPs), appear to be widely disseminated in biological systems. In collaboration with the University of Nottingham, we have identified CDPs as components of stationary phase supernatants from *P. aeruginosa* which demonstrate activity in AHL based bioassays (Holden *et al.*, 1999). CDPs have been identified as being produced by a number of marine bacteria such as *V. parahaemolyticus* and *Micrococcus* (Bell *et al.*, 1994; Stierle *et al.*, 1988) and have also been identified in a number of other biological systems, including: mammalian systems as a neurotransmitter (Prasad, 1995), and in many food products including sake and beer (Gautschi and Schmid, 1997; Takahashi *et al.*, 1974).

A more detailed analysis of the cross-talk of CDPs in other AHL bioassays indicates, that just as different furanones have different activities in AHL assays, different CDPs have different activities as well (Holden *et al.*, 1999). For example, the CDPs L-proline-L-methionine, L-proline-L-valine, L-proline-L-tyrosine, D-alanine-L-valine, and L-proline-L-phenylalanine all induce bioluminescence in the μM range, while L-proline-L-alanine and L-proline-L-leucine have no effect on bioluminescence. However, the differences in activities do not follow a definable structure-function relationship with regard to which AHL phenotype a particular CDP will induce. For example, L-proline-L-tyrosine induces bioluminescence as well as pigment production in an *Agrobacterium* reporter yet L-proline-L-valine, which also induces bioluminescence is not active in the *Agrobacterium* assay (Holden *et al.*, 1999). While AHLs are active at nM concentrations, CDPs are active in AHL system in the μM range. To determine if CDPs mediate their effects by binding to the LuxR protein, displacement experiments similar to those with AHLs and furanones were performed. However, unlike the furanones, no significant displacement of the labeled OHHL in the presence of CDP could be detected at the concentrations which induced bioluminescence (Manefield *et al.* unpublished). This would suggest that either CDPs have a binding affinity for the LuxR which is extremely low and they are therefore incapable of competing with AHLs for the binding site or that they act through another regulator or mechanism. It seems unlikely that CDPs are mediating their effects through some non-specific toxicity as to date there is no report that CDPs are toxic to either prokaryotes or eukaryotes; the food industry taste tests CDPs in mg quantities (Gautschi and Schmid, 1997), and the presence of CDPs in mM concentrations does not effect the growth of either *E. coli* or *P. aeruginosa* (Labbate *et al.* unpublished).

The origin of CDPs in bacteria is also unknown. In the Cyanobacteria, cyclic peptides are produced by a series of non-ribosome peptide synthetases which are encoded by large operons consisting of conserved modules, each of which plays a particular role in cyclisation, recruiting, or modification of the various amino acids making up the cyclic peptide. Another possibility is that the CDPs are produced as degradative products of proteins. In Gram-positive bacteria, the small peptide pheromones used to induce the density dependent phenotypes of competence, virulence expression, or antibiotic production are thought to be derived from limited proteolysis of a pre-peptide (Kleerebezem *et al.*, 1997). It is also suggested that a proline residue near the end of a peptide chain might spontaneously cyclize and self-cleave itself from the peptide, resulting in a cyclic dipeptide(Prasad, 1995). The mechanism of production of, and the genes controlled by, CDPs in bacteria is currently under investigation.

As yet, the biological role of CDPs in bacteria has not been elucidated. *P. aeruginosa* is an example of an organism with two functioning signalling systems, *i.e.* AHLs and CDPs, which may cross-talk with each other, although no specific CDP mediated regulation of *P. aeruginosa* genes has been observed. While there has been no demonstration of CDP mediated phenotypes in *P. aeruginosa,* a recently identified and characterised 2-heptyl-3-hydroxy-4-quinolone molecule has been shown to interact with the AHL quorum sensing pathways in that bacterium (Pesci *et al.*, 1999). These data further support the concept that cross-talk between signalling pathways may not be uncommon. In addition, a homologue of the *V. fischeri luxR, sdiA,* has been identified in *E. coli,* and it is proposed to play a role in the regulation of the cell division genes *ftsQZA* (Wang *et al.*, 1991). Moreover, an as yet unidentified extracellular factor(s) produced by stationary phase *E. coli* cells (Garcia-Lara *et al.*, 1996; Sitnikov *et al.*, 1996; Surrette and Bassler 1998) is also involved in the regulation of these same genes and presumably mediate their effects through SdiA (Garcia-Lara *et al.*, 1996; Sitnikov *et al.*, 1996). Our laboratory has tested the effect of CDPs on the regulation of an *ftsQ-lacZ* fusion and found no difference in LacZ activity at the concentrations tested. However, incubation of a *bolA-lacZ* fusion (Bohannon *et al.*, 1991) in the presence of L-proline-L-leucine, L-proline-L-valine, L-proline-L-tyrosine, L-proline-L-alanine did reduce the LacZ activity of this fusion (Holden*et al.*, 1999), indicating that CDPs may play a role in the regulation of cell morphogenesis through regulation of the stationary phase induced morphogene *bolA* (Aldea *et al.*, 1990). In addition, L-proline-L-tyrosine inhibited expression of the stationary phase gene *fic* (Holden*et al.*, 1999) as determined by monitoring LacZ expression in a *fic-lacZ* fusion (Kawamukai *et al.*, 1989).

As described above, *V. harveyi* has two signalling systems, one that is AHL based and a second that is based on an as yet uncharacterised compound. The second signalling system has proven useful for the identification of novel signals in that it appears to be more promiscuous in its interactions with different signals than the AHL system. Bassler *et al.* (1997) have used this system to identify a number of different bacteria including the marine bacteria *V. cholerae, V. parahaemolyticus, V. anguillarum, V. alginolyticus,* and *E. coli* which produce extra-cellular components that activate bioluminescence through the non-AHL pathway in *V. harveyi.* However, with the exception of *V. harveyi,* the phenotypes regulated by these extra-cellular signals are not known. In addition, two other marine vibrios, *V. vulnificus* and *V. angustum* also produce diffusable signalling molecules which are active in the *V. harveyi* bioassay (deNys *et al.*, 1999; McDougald *et al.*, 1998; Srinivasan *et al.*, 1998). For these two bacteria, the signal extracts appear to be involved in the regulation of starvation induced phenotypes. For example, using *lacZ* reporter transposon mutants of *V. angustum* S14, we have identified a number of carbon starvation induced mutants that also respond to the addition of cell free supernatants from *V. angustum* S14

(Srinivasan *et al.,* unpublished). The addition of the extracellular factor(s) have different effects on the various reporter mutants where some mutants are induced in the presence of the supernatant and others show a reduced expression of the *lacZ* (Srinivasan *et al.,* unpublished). This would lend strong support for the specific effects of the extracellular factor(s) on the regulation of genes that are involved in the starvation response. However, the identities of the signal molecules are unknown. In *E. coli,* Withers *et al.,* (1998) demonstrated that an extracellular factor present in conditioned medium was capable of inhibiting the initiation of chromosomal replication. This suggested that one function of a density dependent signal might be to control cell replication and could be important for the regulation of stationary phase. In another study, the non-AHL signal system has been demonstrated to induce several randomly inserted *lacZ* fusions in *E. coli* (Baca-DeLancy *et al.,* 1999). Interestingly, sequence analysis of those fusions indicated that they are all involved in the uptake, synthesis or degradation of amino acids. Thus it would appear that extracellular factors might be important for regulating adaptive responses during unfavourable growth conditions such as low carbon or stationary phase conditions.

Another useful method to identify bacteria with signalling systems is through gene homology either by genome analysis, hybridisation, or by PCR. Use of these methods has led to the identification of a number of homologues of known quorum sensing genes. Most notably, a homologue of the *V. fischeri luxR* gene has been identified, *sdiA,* in both *E. coli* and *S. typhimurium* (Ahmer *et al.,* 1998; Wang *et al.,* 1991). In *E. coli,* this gene is suspected of being involved in the regulation of cell division, while in *S. typhimurium,* the gene is involved in the regulation of genes encoded by the virulence plasmid (Ahmer *et al.,* 1998). Four marine vibrios, *V. vulnificus, V. angustum, V. cholerae,* and *V. parahaemolyticus* all encode homologues of the *V. harveyi luxR* gene. For the latter two species, this gene is involved in the regulation of proteases and capsule production respectively, while for the former two, the phenotype controlled by this gene has not been identified (Jobling and Holmes, 1997; McCarter, 1998; McDougald *et al.* unpublished).

It should be pointed out that the majority of bacteria discussed herein are Gram-negative and this is primarily because of the focus on AHL systems. However, signalling in Gram-positive bacteria is well established and there is every reason to believe that further studies into these bacteria will reveal many more signalling systems as well. In a continued effort to examine new species of marine bacteria that produce or respond to signal molecules and the phenotypes involved in those, examination of a range of marine isolates by our laboratory indicates that at least 10% of species produce AHLs, although the number and types of phenotypes regulated by those signalling molecules is more difficult to determine (Maximilien *et al.,* 1998; England *et al.* unpublished). This percentage of signal producing marine bacteria is probably an underestimate as it is clear that bacteria are not limited to AHLs as signals. Furthermore, the bioassays used and the signal extraction procedures also have their limitations in sensitivity and hence can also lead to an underestimation of the true number of bacteria that produce signalling molecules. There is also a whole other branch in the tree of life, the Archaea, that has yet to be explored and we anticipate these organisms will also produce their own signals as well as possibly having or being able to respond to those systems described here. The work done with *D. pulchra* as an example of signals or signal antagonists derived from a eukaryote suggests there is some benefit to be gained in looking at organisms besides bacteria for signalling molecules or signal antagonists. As mentioned above, preliminary work with species of *Delisea* besides *D. pulchra* are also yielding novel compounds that appear to cross-talk with AHL based signal systems.

Acknowledgements

Research in the author's laboratory was funded by the Australian Research Council and the Centre for Marine Biofouling and Bio-Innovation. We would like the thank the many members of the research team who contributed information, ideas, and discussions for this manuscript including: Tim Charlton, Symon Dworjanyn, Rocky deNys, Dacre England, Maurice Labbate, Diane McDougald, Mike Manefield, and Sujatha Srinivasan. We also wish to thank our collaborators for their input, Naresh Kumar and Roger Read (School of Chemistry, UNSW), Matt Holden (School of Pharmaceutical Sciences, The University of Nottingham), Lone Gram (Danish Fisheries Institute, Danish Technical Institute) and Patric Nilsson (Department of Theoretical Ecology, Lund University).

References

Ahmer, B.M.M., van Reeuwijk, J., Timmers, C.D., Valentine, P.J., and Heffron, F. 1998. *Salmonella typhimurium* encodes an SdiA homolog, a putative quorum sensor of the LuxR family, that regulates genes on the virulence plasmid. J. Bacteriol. 180: 1185-1193.

Albus, A.M., Pesci, E.C., Runyen-Janecky, L.J., West, S.E.H., and Iglewski, B.H. 1997. Vfr controls quorum sensing in *Pseudomonas aeruginosa*. J. Bacteriol. 179: 3928-3935.

Aldea, M., Garrido, T., Pla, J., and Vicente, M. 1990. Division genes in *Escherichia coli* are expressed coordinately to cell septum requirements by gearbox promoters. EMBO J. 9: 3787-3794.

Baca-DeLancy, R.R., South, M.M.T., Ding, X., and Rather, P.N. 1999. *Escherichia coli* genes regulated by cell-to-cell signalling. Proc. Natl. Acad. Sci. USA. 96: 4610-4614.

Bassler, B., Greenberg, A., and Stevens, A. 1997. Cross-species induction of luminescence in the quorum-sensing bacterium *Vibrio harveyi*. J. Bacteriol. 179: 4043-4045.

Bassler, B., and Silverman, M.R. 1995. Intercellular communication in marine *Vibrio* species: Density-dependent regulation of the expression of bioluminescence. In: Two -component signal transduction, Hoch, J. A., and Silhavy, T.J., eds., ASM Press, Washington. p. 431-444.

Bassler, B.L., Wright, M., and Silverman, M.R. 1994. Multiple signalling systems controlling expression of luminescence in *Vibrio harveyi* sequence and function of genes encoding a second sensory pathway. Mol. Microbiol. 13: 273-286.

Bell, R., Carmeli, S., and Sar, N. 1994. Vibrindole A. a metabolite of the marine bacterium, *V. parahaemolyticus*, isolated from the toxic mucus of the boxfish *Ostracion cubicus*. J. Nat. Prod. 57: 1587-1590.

Bohannon, D.E., Connell, N., Kleener, J., Tormo, A., Espinosa-Urgel, M., Zambrano, M., and Kolter, R. 1991. Stationary-phase inducible "Gearbox" promoters: Differential effects of *katF* mutations and the role of sigma 70. J. Bacteriol. 173: 4482-4492.

Cha, C., Gao, P., Chen, Y., Shaw, P., and Farrand, S. 1998. Production of acyl-homoserine lactone quorum-sensing signals by gram-negative plant-associated bacteria. Mol. Plant Microbe Interactions. 11: 1119-1129.

Clewell, D.B., and Weaver, K.E. 1993. Sex pheromones and plasmid transfer in *Enterococcus faecalis*. Plasmid. 21: 175-184.

Davies, D.G., Parsek, M.R., Pearson, J.P., Iglewski, B.H., Costerton, J.W., and Greenberg, E.P. 1998. The involvement of cell-cell signals in the development of a bacterial biofilm. Science. 280: 295-298.

deNys, R., Dworjanyn, S.A., and Steinberg, P.D. 1998. A new method for determining surface concentrations of marine natural products on seaweeds. Mar. Ecol. Prog. Ser. 162: 79-87.

deNys, R., Rice, S.A., Manefield, M., Srinivasan, S., McDougald, D., Loh, A., Ostling, J., Lindum, P., Givskov, M., Steinberg, P., and Kjelleberg, S. 1999. Cross-talk in bacterial extracellular signalling systems. In: Microbial Biosystems: New Frontiers. Proceedings of the 8th Intl. Symp. on Microbial Ecology., Bell, C. R., Brylinsky, M., and Johnson-Green, P., eds., Atlantic Canada Society for Microbial Ecology, Halifax. (In press).

deNys, R., Wright, A.D., Konig, G.M., and Sticher, O. 1993. New halogenated furanones from the marine alga *Delisea pulchra* (cf fimbriata). Tetrahedron. 49: 11213-11220.

Eberhard, A.A., Burlingamme, L., Eberhard, C., Kenyon, G.L., and Nealson, K.H. 1981. Structural identification of autoinducer of *Photobacterium fischeri* luciferase. Biochem. 20: 2444-2449.

Eberl, L., Winson, M., Sternberg, C., Stewart, G., Christense, G., Chhabra, S., Bycroft, P., Williams, S., Molin, S., and Givskov, M. 1996. Involvement of N-acyl-homoserine lactone autoinducers in controlling the multicellular behaviour of *Serratia liquefaciens*. Mol. Microbiol. 20: 1127-1136.

Evans, K., Passador, L., Srikumar, R., Tsang, E., Nezezon, J., and Poole, K. 1998. Influence of the MexAB-OprM multidrug efflux system on quorum sensing in *Pseudomonas aeruginosa*. J. Bacteriol. 180: 5443-5447.

Flavier, A.B., Clough, S. J., Schell, M.A., and Denny, T.P. 1997. Identification of 3-Hydroxypalmitic acid methyl ester as a novel autoregulator controlling virulence in *Ralstonia solanacearum*. Mol. Microbiol. 26: 251-259.

Fuqua, C., Winans, S.C., and Greenberg, E.P. 1996. Census and Consensus in bacterial ecosystems: The LuxR-LuxI family of quorum-sensing transcriptional regulators. Ann. Rev. Microbiol. 50: 727-751.

Fuqua, W.C., Winans, S.C., and Greenberg, E.P. 1994. Quorum sensing in bacteria: The LuxR-LuxI family of cell density-responsive transcriptoinal regulators. J. Bacteriol. 176: 269-275.

Garcia-Lara, J., Shang, L.H., and Rothfield, L.I. 1996. An extracellular factor regulates expression of *sdiA*, a transcriptional activator of cell division genes in *Escherichia coli*. J. Bacteriol. 178: 2742-2748.

Gautschi, M., and Schmid, J.P. 1997. Chemical characterization of diketopiperiazines in beer. J. Agric. Food Chem. 45: 3183-3189.

Givskov, M., deNys, R. Manefield, M., Gram, L., Maximilien, R., Eberl, L., Molin, S., Steinberg, P., Kjelleberg, S. 1996. Eukaryotic interference with homoserine lactone-mediated prokaryotic signalling. J. Bacteriol. 178:6618-6622.

Gram, L., Christensen, A.B., Ravn, L., Molin, S., Givskov, M. 1999. Production of acylated homoserine lactones by psychrotrophic members of the *Enterobacteriaceae* isolated from foods. Appl. Environ. Microbiol. 65: 3458-3463.

Holden, M.T.G., Chahabra, S.R., deNys, R., Stead, P., Bainton, N.J., Hill, P.J., Manefield, M., Kumar, N., Labbate, M., England, D., Rice, S.A., Givskov, M., Salmond, G., Stewart, S.A.B., Bycroft, B.W., Kjelleberg, S., and Williams, P. 1999. Quorum sensing cross talk: Isolation and chemical characterisation of cyclic dipeptides from *Pseudomonas aeruginosa* and other Gram-negative bacteria. Mol. Microbiol. 33: 1254-1266.

Huang, J., Yindeeyoungyeon, W., Garg, R.P., Denny, T.P., and Schell, M. A. 1998. Joint transcriptional control of *xpsR*, the unusual signal integrator of the *Ralsotonia solancearum* virulence gene regulatory network, by a response regulator and a LysR-type transcriptional activator. J. Bacteriol. 180: 2736-2743.

Jobling, M.G., and Holmes, R.K. 1997. Characterization of *hap*R, a positive regulator of the *Vibrio cholerae* HA/protease gene *hap*, and its identification as a functional homologue of the *Vibrio harveyi lux*R gene. Mol. Microbiol. 26: 1023-1034.

Kaplan, H., and Greenberg, E.P. 1985. Diffusion of autoinducer is involved in regulation of the *Vibrio fischeri* luminescence system. J. Bacteriol. 163: 1210-1214.

Kawamukai, M., Matsuda, H., Fujii, W., Utsumi, R., and Komano, T. 1989. Nucleotide sequences of *fic* and *fic-1* genes involved in cell filamentation induced by cyclic AMP in *Esherichia coli*. J. Bacteriol. 171: 4525-4529.

Kazlauskas, R., Murphy, P.T., Quinn, R.J., and Wells, R.J. 1977. A new class of halogenated lactones from the red alga *Delisea fimbriata* (Bonnemaisoniaceae). Tetrahedron Lett. 1: 37-40.

Kitten, T., Kinscherf, T.G., Mcevoy, J.L., and Willis, D.K. 1998. A newly identified regulator is required for virulence and toxin production in *Pseudomonas syringae*. Mol. Microbiol. 28: 917-929.

Kjelleberg, S., Steinberg, P., Givskov, M., Gram, L., Manefield, M., and Denys, R. 1997. Do marine natural products interfere with prokaryotic AHL regulatory systems. Aquatic Microbial Ecology. 13: 85-93.

Kleerebezem, M., Quadri, L.E.N., Kuipers, O.P., and deVos, W.M. 1997. Quorum sensing by peptide pheromones and two component signal transduction systems in Gram positive bacteria. Mol. Microbiol. 24: 895-904.

Kuspa, A., Plamann, L., and Kaiser, D. 1992a. A-signaling and the cell density requirement for *Myxococcus xanthus* development. J. Bacteriol. 174: 7360-7369.

Kuspa, A., Plamann, L., and Kaiser, D. 1992b. Identification of heat-stable A-factor from *Myxococcus xanthus*. J. Bacteriol. 174: 3319-3326.

Latifi, A., Winson, M.K., Foglino, M., Bycroft, B.W., Stewart, G.S.A.B.,

Lazdunski, A., and Williams, P. 1995. Multiple homologues of LuxR and LuxI control expression of virulence determinants and secondary metabolites through quorum sensing in *Pseudomonas aeruginosa* PA01. Mol. Microbiol. 17: 333-343.

Latifi, M.F., Tanaka, K., Williams, P., and Lazdunski, A. 1996. A hierarchial quorum-sensing cascade in *Pseudomonas aeruginosa* links the transcriptional activators LasR and RhlR (VsmR) to expression of the stationary phase sigma factor RpoS. Mol. Microbiol. 21: 1137-1146.

Manefield, M., deNys, R., Kumar, N., Read, R., Givskov, M., Steinberg, P., and Kjelleberg, S. 1999. Evidence that halogenated furanones from *Delisea pulchra* inhibit AHL mediated gene expression by displacing the AHL signal from its receptor protein. Microbiol. 145:283-291.

Maximilien, R., deNys, R., Holmstrom, C., Gram, L., Givskov, M., Crass, K., Kjelleberg, S., and Steinberg, P. 1998. Chemical mediation of bacterial surface colonisation by secondary metabolites from the red alga *Delisea pulchra*. Aquatic Microb. Ecol. 15: 233-246.

McCarter, L.L. 1998. OpaR, a homolog of *Vibrio harveyi* LuxR, controls opacity of *Vibrio parahaemolyticus*. J. Bacteriol. 180: 3166-3173.

McClean, K.H., Winson, M.K., Fish, L., Taylor, A., Chhabra, S., Camara, M., Daykin, M., Lamb, J.H., Swift, S., Bycroft, B.W., Stewart, G.S.A.B., and Williams, P. 1997a. Quorum-sensing and *Chromobacterium violaceum*: Exploitation of violacein production and inhibition for the detection of N-acylhomoserine lactones. Microbiol. 143: 3703-3711.

McClean, R.J.C., Whiteley, M., Stickler, D.J., and Fuqua, W.C. 1997b. Evidence of autoinducer activity in naturally occurring biofilms. FEMS Micrbiol. Lett. 154: 259-263.

McDougald, D., Rice, S.A., Weichart, D., and Kjelleberg, S. 1998. Non-culturability: Adaptation or debilitation. FEMS Microbiol. Ecol. 25: 1-9.

Milton, D.L., Hardman, A., Camara, M., Chhabra, S.R., Bycroft, B.W., Stewart, G., and Williams, P. 1997. Quorum sensing in *Vibrio anguillarum* characterization of the *vanI/vanR* locus and identification of the autoinducer N (3 oxodecanoyl) homoserine lactone. J. Bacteriol. 179: 3004-3012.

Pesci, E.C., Milbank, J.B.J., Pearson, J.P., McKnight, S., Kende, A.S., Greenberg, E.P., and Iglewski, B.H. 1999. Quinolone signalling in the cell-to-cell communication system of *Pseudomonas aeruginosa*. Proc. Natl. Acad. Sci. USA. 96: 11229-11234.

Piper, K.R., Beck von Bodman, S., and Farrand, S.K. 1993. Conjugation factor of *Agrobacterium tumefaciens* regulates Ti plasmid transfer by autoinduction. Nature. 362: 448-450.

Prasad, C. 1995. Bioactive cyclic peptides. Peptides. 16: 151-164.

Shaw, P.D., Ping, G., Daly, S.L., Cha, C., Cronan, J.E., Rinehart, K.L., and Farrand, S.K. 1997. Detecting and characterizing *N*-acyl homoserine lactone signal molecules by thin layer chromatography. Proc. Natl. Acad. Sci. USA. 94: 6036-6041.

Sitnikov, D.M., Schineller, J.B., and Baldwin, T.O. 1996. Control of cell division in *Eshcerichia coli*: Regulation of transcription of *ftsQA* involves both *rpoS* and SdiA-mediated autoinduction. Proc. Natl. Acad. Sci. USA. 93: 336-341.

Srinivasan, S., Ostling, J., Charlton, T., deNys, R., Takayama, K., and

Kjelleberg, S. 1998. Extra-cellular signal molecule(s) involved in the carbon starvation response of marine *Vibrio* sp. strain S14. J. Bacteriol. 180: 201-209.

Stickler, D.J., Morris, N.S., McLean, R.J.C., and Fuqua, C. 1998. Biofilms on indwelling urethral catheters produce quorum-sensing signal molecules *in situ* and *in vitro*. Appl. Environ. Microbiol. 64: 3486-3490.

Stierle, A.C., Cardellina, J.H., and Singleton, F.L. 1988. A marine *Micrococcus* produces metabolites ascribed to the sponge *Tedania ignis*. Experimentia. 44: 1021.

Surrette, M., and Bassler, B. 1998. Quorum sensing in *Escherichia coli* and *Salmonella typhimurium*. Proc. Natl. Acad. Sci. USA. 95: 7046-7050.

Surette, M.G., Miller, M.B., and Bassler, B.L. 1999. Quorum sensing in *Escherichia coli, Salmonella typhimurium* and *Vibrio harveyi:* A new family of genes responsible for autoinducer production. Proc. Natl. Acad. Sci. USA. 96: 1639-1644.

Swift, S., Karlyshev, A.V., Fish, L., Durant, E.L., Winson, M.K., Chhabra, S. R., Williams, P., Macintyre, S., and Stewart, G. 1997. Quorum sensing in *Aeromonas hydrophila* and *Aeromonas salmonicida* - identification of the *lux*RI homologs *ahy*RI and *asa*RI and their cognate N-acylhomoserine lactone signal molecules. J. Bacteriol. 179: 5271-5281.

Swift, S., Throup, J.P., Williams, P., and Stewart, G.S.A.B. 1996. Quorum sensing: a population-density component in the determination of bacterial phenotype. Trends Biochem. Sci. 21: 214-219.

Takahashi, K., Tadenuma, M., Kitamoto, K., and Sato, S. 1974. L-Prolyl-L-leucine anhydride: A bitter compound formed in aged sake. Agric. Biol. Chem. 38: 927-938.

Telford, G., Wheeler, D., Williams, P., Tomkins, P.T., Appleby, P., Sewell, H., Stewart, G.S.A.B., Bycroft, B.W., and Pritchard, D. 1998. The *Pseudomonas aeruginosa* quorum sensing singnal molecule N-(3-Oxydodecanoyl)-L-homoserine lactone has immunomodulatory activity. Infect. Immun. 66: 36-42.

Ulitzer, S., and Kuhn, J. 1988. The transcription of bacterial luminescence is regulated by sigma 32. J. Biolum. Chemilum. 2: 81-93.

Ulitzur, S., and Dunlap, P.V. 1995. Regulatory circuitry controlling luminescence autoinduction in *Vibrio fischeri*. Photochem. Photobiol. 62: 625-632.

Ulitzur, S., Matin, A., Fraley, C., and Meighen, E. 1997. HNS protein represses transcription of the *lux* systems of *Vibrio fischeri* and other luminous bacteria cloned into *Escherichia coli*. Curr. Microbiol. 35: 336-342.

Van Haastert, P.J.M. 1991. Transmembrane signal transduction pathways in *Dicyostelium*. In: Advances in second messenger and phosphoprotein research, Greengard, P., and Robinson, G. A., eds., Raven Press, Ltd., New York. p. 185-226.

Wang, X., de Boer, P.A.J., and Rothfield, L.I. 1991. A factor that positively regulates cell division by activating transcription of the major cluster of essential cell division genes of *Escherichia coli*. EMBO J. 10: 3363-3372.

Winson, M.K., Swift, S., Fish, L., Throup, J.P., Jorgensen, F., Chhabra, S. R., Bycroft, B.W., Williams, P., and Stewart, G. 1998. Construction and analysis of *luxCDABE*-based plasmid sensors for investigating N-Acyl homoserine lactone-mediated quorum sensing. FEMS Microbiol. Lett. 163: 185-192.

Withers, H.L., and Nordstrom, K. 1998. Quorum-sensing acts at initiation of chromosomal replication in *Escherichia coli.* Proc. Natl. Acad. Sci. USA. 95: 15694-15699.

Yamada, Y., Sugamura, K., Kondo, K., Yanagimoto, M., and Okada, H. 1987. The structure of inducing factors for viginiamycin production in *Streptomyces virginiae.* J. Antibiot. 40: 496-504.

Zhang, L.H., Murphy, P.J., Kerr, A., and Tate, M.E. 1993. *Agrobacterium* conjugation and gene regulation by *N*-acyl-L-homserine lactones. Nature. 362: 446-448.

5

Microbial Symbionts of Marine Invertebrates: Opportunities for Microbial Biotechnology

Margo G. Haygood, Eric W. Schmidt, Seana K. Davidson, and D. John Faulkner

Marine Biology Research Division and Center for Marine Biotechnology and Biomedicine, Scripps Institution of Oceanography, University of California, San Diego, La Jolla, CA 92093-0202, USA

Abstract

Marine invertebrates are sources of a diverse array of bioactive metabolites with great potential for development as drugs and research tools. In many cases, microorganisms are known or suspected to be the biosynthetic source of marine invertebrate natural products. The application of molecular microbiology to the study of these relationships will contribute to basic biological knowledge and facilitate biotechnol-ogical development of these valuable resources. The bryostatin-producing bryozoan *B. neritina* and its specific symbiont "*Candidatus* Endobugula sertula" constitute one promising model system. Another fertile subject for investigation is the listhistid sponges that contain numerous bioactive metabolites, some of which originate from bacterial symbionts.

Introduction

Most of the metabolic and biochemical diversity of life resides in microorganisms: the domains Bacteria and Archaea, and unicellular members of the Eukarya. Eukaryotic multicellular organisms, although more conspicuous, are limited in their biochemical capabilities. Thus, it is not surprising that microbial symbionts have been adopted by higher organisms time and again throughout evolution. Indeed, the eukaryotes are the product of symbiotic associations between early eukaryotic cells and bacteria that became mitochondria and chloroplasts, greatly expanding the ecological opportunities for these chimeric organisms.

Both the fact and the function of symbiosis are often relatively obvious. For example, many fishes and some squid maintain cultures of bioluminescent bacteria whose light is put to a multitude of uses (Haygood, 1993; McFall-Ngai, 1994). The hydrothermal vent vestimentiferans and bivalves that are supported by chemoautotrophic symbiotic bacteria are noteworthy examples that were readily recognized because their unique habitat and unusual morphology pointed inevitably to dependence on symbionts (Cavanaugh, 1994). Likewise, the dependence of many cnidarians such as corals on photosynthetic symbionts is well known (Buddemeier, 1994; Rowan, 1998).

It is becoming apparent, however, that symbionts can fulfill much more diverse and subtle roles. In particular, animals can gain access to the biosynthetic virtuosity of prokaryotes through symbiosis. Marine invertebrates, particularly sessile ones, are rich sources of unusual metabolites. Like terrestrial plants, they often rely upon chemical defense to discourage predation. Symbiosis with biochemically versatile microorganisms is an efficient strategy to accomplish chemical defense. Microbial symbionts have often been invoked as biosynthetic sources for natural products found in marine invertebrates. In some cases, the evidence supports an autogenous source for the compounds, while in others the data support a microbial symbiont origin. The vast majority lies in a gray area, where arguments based on structural similarity to microbial metabolites often suggest a symbiotic origin, but solid evidence is lacking. There are difficulties in relying upon structural similarities to speculate about the source of compounds. Similar and identical compounds can arise from parallel or convergent evolution. Similarities can also arise from digestive degradation of standard compounds, such as the dioxopiperazines formed by hydrolysis of peptides and proteins. Another potential problem is transfer of genes from bacteria to other organisms (Hopwood, 1997). Genes for many biosynthetic pathways are contiguous in bacteria, and it is reasonable to assume that pieces of these genes, as well as entire gene clusters, may be moved by lateral gene transfer. Thus, specialized genes encoding unusual functions could be obtained by gene transfer from bacteria to invertebrates rather than evolving *de novo*. A high frequency of plasmid transfer in the marine environment amplifies the possibility of lateral gene transfer (Dahlberg *et al.*, 1998). Finally, it is premature to declare that because a compound has only previously been found in a single source such as a species of microbe, related compounds must come from the same microbe. It is equally possible that other sources have not yet been found. Hypotheses concerning the biosynthetic source for a compound must be investigated in each instance.

Natural products from marine invertebrates greatly expand the chemical diversity available for biotechnological exploitation. The soil microbiota and terrestrial plants have proven to be extraordinary repositories of diverse compounds that can be employed directly or modified for application. The marine realm promises to be another excellent source with little overlap with traditional sources of natural products. Developing these resources can present problems, however. The organisms can be rare, slow growing or difficult to collect. Wild collection can cause environmental damage to vulnerable habitats. Aquaculture is an option in some cases, but even aquaculture can have negative environmental consequences.

Marine invertebrate natural product symbioses present a fascinating subject for basic biological and biochemical research. In addition, they present an opportunity to develop these resources in ways that circumvent environmental and supply problems. Cultivating the symbionts, understanding the regulation of compound production and optimizing biosynthesis *ex symbio* completely eliminates the problems of wild collection, and can potentially improve the yield and thus the economics of production. However, cultivation of microbial symbionts is not a trivial problem.

The major challenge in microbial ecology today is that the microbes that predominate in natural environments are not well represented among laboratory strains (Hugenholtz *et al.*, 1998). We need to develop more insightful methods for making these organisms amenable to laboratory investigation. Symbioses are ideal systems for addressing these problems because the populations present are relatively stable, and the constituent organisms range in ease of cultivation. We need to expand our thinking beyond pure cultures that grow rapidly to high cell densities to include consortia and growth conditions that more closely mimic *in situ* conditions. We can obtain sequence information from ribosomal RNA (rRNA) genes directly from the symbiosis without cultivation, and design specific oligonucleotide probes from these sequences. With these tools we can monitor symbionts of interest in consortia and dissect their function in a way that was previously impossible.

Symbiotic systems are also well suited to direct investigation of biosynthesis using molecular biological techniques. Using database sequences as a starting point, we can clone biosynthetic genes directly from the symbiosis and reconstitute them in a heterologous host for expression and compound production. The fact that many bacterial pathways for secondary metabolites are organized in contiguous operons facilitates pathway reconstruction. Once the genes are in hand, combinatorial or pathway engineering approaches can be used to obtain novel structures with different activities.

Examples

Microbial symbioses have been described from most phyla of marine invertebrates: Porifera, Cnidaria, Bryozoa, Mollusca, Pogonophora, Echinodermata, Urochordata and Crustacea. Symbionts include bacteria, archaea and unicellular eukaryotes such as dinoflagellates. Interesting natural products have been found particularly in sponges, ascidians, cnidarians, bryozoans, and nudibranchs (Anthoni *et al.*, 1990). In this article, we will describe examples of both simple and complex symbiotic systems, focusing on bryozoans and sponges.

Bryozoans

Bryozoans are small, sessile colonial invertebrates. Bacterial symbionts have been discovered in several species of bryozoans, but most have not been examined for symbionts (Lutaud, 1964; 1965; 1969; 1986; Woollacott and Zimmer, 1975; Woollacott, 1981; Zimmer and Woollacott, 1983; Boyle *et al.*,

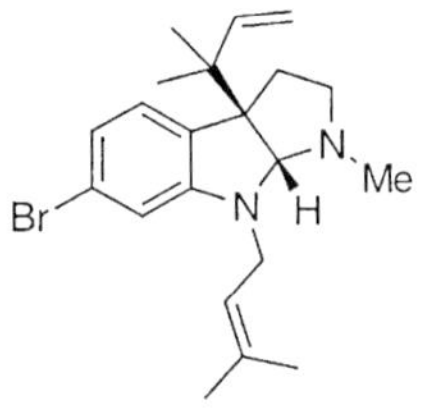

Flustramine A

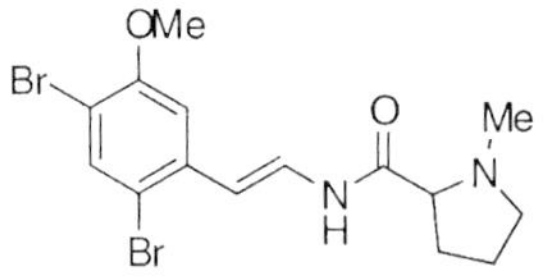

Amathamide A

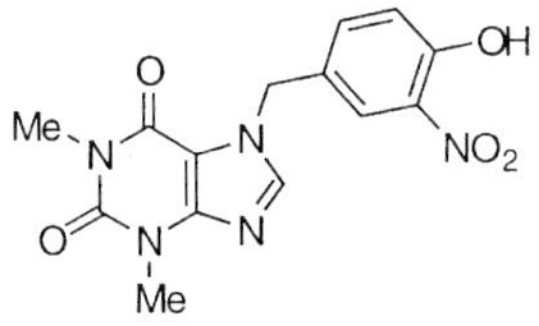

Phidolopin

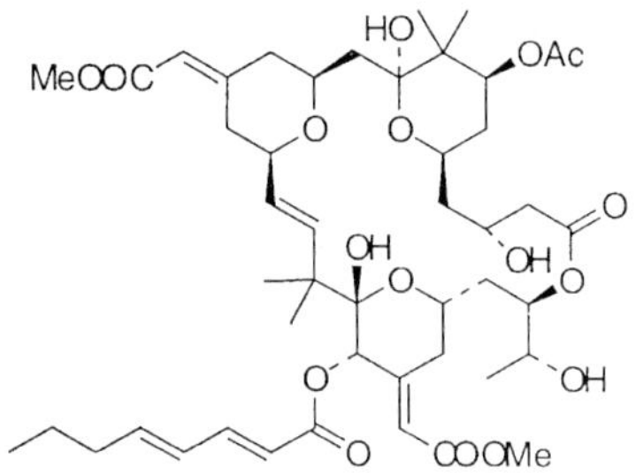

Bryostatin 1

Figure 1. Bryozoan Natural Products

1987). The natural products most commonly found in bryozoans are alkaloids. In addition to alkaloids, bryozoans contain other classes of compounds. *Watersipora cucullata* contains sulfated ceramides (Ojika *et al.*, 1997), *Dakaria subovoidea*, contains a thiophene (Shindo *et al.*, 1993), *Alcyonidium gelatinosum* contains dimethyl sulfoxonium ion (Christophersen, 1985), and a family of macrocylic polyketides, the bryostatins, are found in *B. neritina* (Pettit, 1991). A symbiotic origin has been suggested for several bryozoan natural products: alkaloids such as the flustramines and the amathamides, as well as the phidolopins and the bryostatins (Anthoni *et al.*, 1990) (Figure 1). Bryozoan pyrrrole pigments such as the tambjamines are related to bacterial products and thus could be of dietary or symbiotic origin (Anthoni *et al.*, 1990).

Bacterial Symbionts in Bryozoans

In adult bryozoan colonies, bacterial symbionts are typically associated with the funicular cords, which are internal organs that connect individuals of the colonies (Lutaud, 1969; Woollacott and Zimmer, 1975; Zimmer and Woollacott, 1983). In some species, they reside in other structures apparently specialized for housing symbionts (Lutaud, 1964; 1965; 1986). In a few cases, bacteria have been found in modified tissues in the larvae (Woollacott, 1981; Zimmer and Woollacott, 1983). Such adaptations to ensure intergenerational transmission suggest a highly evolved association in which the presence of the bacteria is advantageous to the bryozoan. Bryozoans typically have a simple associated microbial community relative to sponges. This should make bryozoans relatively tractable model systems for the investigation of symbiosis.

Bugula neritina

B. neritina is the source of the bryostatins, a family of cytotoxic macrolides (Pettit, 1991). B. neritina is a case in which a symbiotic origin of the bryostatins has been proposed, but not yet proven.

B. neritina is a cosmopolitan, temperate fouling invertebrate often found on boats, piers and subtidal rocky substrates. It is considered a "weed" that grows rapidly and outcompetes the other fouling organisms. The feeding zooids extend a crown of tentacles called a lophophore that captures phytoplankton and other small particles. B. neritina broods its larvae in specialized zooids called ovicells (Woollacott and Zimmer, 1975). The larvae are released when mature and competent to settle. The larvae are relatively large (approx. 200-300 µm). They have no gut, and do not feed. Upon settling they rapidly metamorphose and produce the ancestrula, the first colonial unit with a feeding zooid (Woollacott, 1971; Zimmer and Woollacott, 1977; Woollacott and Zimmer, 1978). The larvae typically settle near other B. neritina, probably due to limited dispersal (Keough, 1989). B. neritina can be raised from larvae to reproductive adults in the laboratory, although growth rates are generally lower than in wild colonies. B. neritina larvae contain bryostatins and post-spawn colonies have less bryostatin (Thompson and Mendola, unpublished). Bryostatins appear to be stable and persistent in the colony. The highest bryostatin contents we have observed occurred in very old colonies at a site protected from winter storms, suggesting that it accumulates over time. Even completely senescent colonies that have been reduced to skeletons contain significant quantities of bryostatins, perhaps due to diffusion of bryostatins throughout the colony. Bryostatins do not originate in the diet of B. neritina since colonies raised on a completely artificial, bryostatin-free diet produce bryostatins (Thompson and Mendola, unpublished).

Chemical defense of B. neritina has not been extensively studied. A study of chemical defense of invertebrate larvae found that the larvae of B. neritina are strongly chemically defended, but did not find evidence for defense in the adult colony (Lindquist and Hay, 1996). It is possible that bryostatins are produced by the adult colony and sequestered in the larvae to defend the larvae.

Bacterial Symbionts of B. neritina

The larvae of B. neritina contain rod-shaped Gram-negative bacteria in the pallial sinus, a sealed circular fold in the aboral surface of the larva (Woollacott, 1981). The bacteria appear to be extracellular by transmission electron microscopy. B. pacifica and B. simplex contain morphologically similar bacteria but B. turrita and B. stolonifera do not (Woollacott, 1981). In B. neritina, the bacteria can be seen in the pallial epithelium during settling and metamorphosis (Reed and Woollacott, 1983). In adult colonies, bacteria occur in the funicular cords that connect zooids and nourish the developing larvae (Woollacott and Zimmer, 1975). Whether these bacteria are the same as those in the larvae is not known, but if they are, their presence in the funicular cords would provide a mechanism for them to be transferred to the larvae and to regenerating zooids. The larval symbiont is a candidate for a microbial source of bryostatins. The elaborate adaptations to ensure transmission of symbionts from adults to new colonies could account for the universality of bryostatin production in *B. neritina.*

   B. neritina is the only bryozoan for which symbiont rRNA sequence data is available (Haygood and Davidson, 1997). Small subunit (SSU, 16S) rRNA genes of the symbionts were amplified from DNA extracted from washed larvae using universal and eubacterial primers and directly sequenced. Specific oligonucleotides were synthesized based on the sequence and one was used as a probe for *in situ* hybridization. The specific probe binds to the bacteria in the pallial sinus but not to other related bacteria, confirming the authenticity of the sequence. Based on *in situ* hybridization, there are about 2500 pallial sinus symbionts per larva. Analysis of sequences from larvae from seven different West Coast populations and one from the Atlantic Ocean shows that the same symbiont is present in all cases. Amplification with specific oligonucleotides confirms the presence of this organism in all 16 populations tested. There are two strains that differ at four nucleotide sites, discussed below. Phylogenetic analysis of the sequences shows that the symbiont is a γ-proteobacterium that is not closely related to any sequence in the databases. This organism has been named " *Candidatus* Endobugula sertula" (abbreviated *"E. sertula"*). This taxonomic status is frequently used for bacteria for which a type strain is not available (Murray and Stackebrandt, 1995).

   A highly specific and sensitive PCR assay for the presence of the symbionts was developed using the specific oligonucleotides. The larvae were tested to see if *"E. sertula"* could be readily cultivated by plating larval samples on standard media and then testing the strains that grew with the PCR assay. None of these strains proved to be *"E. sertula"*, demonstrating that it will require a specialized method for cultivation. Bacterial DNA from seawater near *B. neritina* colonies was negative, suggesting that the symbiont is not a major component of free-living bacterial populations.

Bryostatins
Bryostatins are an important family of cytotoxic macrolides based on the bryopyran ring system (Pettit, 1991).They occur exclusively in *B. neritina.* Bryostatin 1 is now in Phase II clinical trials for treatment of leukemias, lymphomas, melanoma and solid tumors (Pluda *et al.*, 1996). Bryostatin 1 also shows promise for treatment of ovarian and breast cancer and to enhance lymphocyte survival during radiation treatment (Kraft, 1993; Lind *et al.*, 1993; Grant *et al.*, 1994; Scheid *et al.*, 1994; Sung *et al.*, 1994; Correale *et al.*, 1995; Fleming *et al.*, 1995; Baldwin *et al.*, 1997; Lipshy *et al.*, 1997; Taylor *et al.*, 1997).

   Unlike most chemotherapeutic agents that kill rapidly dividing cells, bryostatins act on signal transduction pathways by binding to the activator site of protein kinase C (Steube and Drexler, 1993; Caponigro *et al.*, 1997). Eighteen bryostatins have been described (Pettit *et al.*, 1982; 1991; 1996). These vary primarily in the substituents at C-7 and C-20. Incorporation of the octa-2,4-dienoate ester at C-20 is a unique capability of *B. neritina* from California (Pettit, 1991). Two chemotypes have been defined as follows (Davidson and Haygood, 1999). Chemotype O refers to samples that contain the octa-2,4-dienoate ester at C-20 (bryostatins 1-3, 12 and 15) as well as other bryostatins. Chemotype M lacks the octa-2,4-dienoate ester at C-20, and the presence of other bryostatins in the absence of bryostatins 1-3 is diagnostic of this chemotype. The compound in clinical trials, bryostatin 1, is only found in chemotype O and this chemotype will be the target of commercial exploitation.

The major obstacle in investigating and developing bryostatins as anti-cancer agents or for other therapeutic purposes is the difficulty of obtaining them in ample quantities. The yield of bryostatin 1 is low; in the large-scale isolation for clinical trials it was 1.4 µg per gram wet weight of *B. neritina*. (Schaufelberger *et al.*, 1991). The supply of *B. neritina* is unpredictable and harvesting has long-term negative effects on populations. The population collected for production of bryostatin 1 for clinical trials took several years to regrow. A recent study (Davidson and Haygood, 1999) showed that there are two genetic types (probably species) of *B. neritina*. Type D (deep) is found in waters deeper than 9 m in Southern California, has one strain of "*E. sertula*" (designated type D) and has chemotype O bryostatins, including bryostatin 1. Type S (shallow) is found at less than 9 m and also occurs in the Atlantic. It has another strain of "E. sertula" (S), and has only chemotype M bryostatins, lacking bryostatin 1. The S type of *B. neritina* is transported on boat hulls and may occur worldwide. It is not known whether the difference in chemotype is due to the genetic difference in *B. neritina* or the different strains of "*E. sertula*". The type D population containing bryostatin 1 is more limited than was previously appreciated. If the clinical trials succeed, supply from nature will be a serious problem for commercial production of bryostatin 1.

Research has focused on bryostatin 1, but the other bryostatins have been isolated on the basis of their antileukemic activity. With the exception of bryostatins 16 and 17, all possess the structural features believed to account for the activity of bryostatin 1 (Pettit *et al.*, 1982; 1991; 1996). Other bryostatins may equal or exceed the therapeutic value of bryostatin 1. One study showed that bryostatins 5 and 8 are as effective as bryostatin 1 in treating melanoma, but with milder side effects (Kraft *et al.*, 1996). Greater availability of other bryostatins is essential to permit research to unlock the potential of this remarkable family of compounds. Although aquaculture would provide a more consistent source of bryostatins, it does not improve the low yield. Chemical synthesis of bryostatins is currently considered impractical due to their structural complexity. However, synthetic analogs have promise (Kageyama *et al.*, 1990; Wender *et al.*, 1998).

Biosynthesis of Bryostatins
Polyketides are created by sequential condensation of acetate or other simple fatty acid units as in fatty acid synthesis (Katz and Donadio, 1993). There are two types of cyclic polyketides, complex and aromatic. Bryostatins are complex polyketides.

Complex polyketides are typically microbial secondary metabolites. This raises the possibility that bryostatins are made by bacteria associated with *B. neritina* rather than by *B. neritina* itself (Anthoni *et al.*, 1990). Complex polyketides are best known in the actinomycetes, which are members of the high G+C Gram positive bacteria. Microorganisms other than actinomycetes that produce complex polyketides include cyanobacteria, mycobacteria (high G+C Gram positive), myxobacteria (delta proteobacteria), pseudomonads (γ-proteobacteria) and fungi (Katz and Donadio, 1993; Schupp *et al.*, 1995; Nowak-Thompson *et al.*, 1997).

Although complex polyketides are generally regarded as microbial secondary metabolites, some have been isolated from metazoans. Miyakolide, swinholide A, halichondrin, discodermolide and spongistatin are complex polyketides isolated from sponges (Uemura *et al.*, 1985; Hirata and Uemura, 1986; Kobayashi *et al.*, 1990; Higa *et al.*, 1992; Bai *et al.*, 1993; Pettit *et al.*, 1993; Litaudon *et al.*, 1994). However, sponges harbor many bacteria that could produce these compounds, as discussed below.

The enzymes responsible for biosynthesis of polyketides, polyketide synthases (PKS), and fatty acid synthases (FAS) are categorized as Type I, with multiple active sites on a single polypeptide, or Type II, with single active site polypeptides that form a complex. Bacterial FAS is Type II, but eukaryotic FAS is Type I. In both bacteria and fungi, complex polyketides are made only by Type I PKS (PKS-I). We postulate that the PKS responsible for synthesis of the bryopyran ring (bryopyran synthase) is a PKS-I. Since no metazoan enzymes responsible for synthesis of complex polyketides are known, we should consider the possibility is that the bryopyran synthase is a microbial enzyme.

A study of incorporation of radioactively labeled biosynthetic building blocks by *B. neritina* demonstrated that acetate is the probable precursor of the bryopyran ring, with methyl groups on the macrolactone ring system added by *S*-adenosyl methionine after polyketide chain assembly (Kerr *et al.*, 1996). These results are consistent with a polyketide synthase mechanism.

B. neritina as a Model System

It will be worthwhile to investigate whether the biosynthetic source of bryostatins is a microorganism, particularly "*E. sertula*". Cultivation of such a microorganism, or cloning of the genes, could provide a solution to the problem of supply of bryostatins.

In addition, the *B. neritina*/ "*E. sertula*" system has features that are favorable for development as a model system for the study of natural product symbioses using microbiological and molecular approaches. A good model system is tractable and accessible, yet sufficiently representative to yield useful knowledge applicable to other systems. The most important advantage of this system is simplicity of the microbial community consisting of the single bacterial symbiont and relatively few other bacteria, in contrast to the complex mutualistic, commensal, and ingested microbial populations of most sponges. In *B. neritina*, symbiont ribosomal sequences have been obtained, specific assay methods have been developed and extensive surveys carried out that demonstrate that "*E. sertula*" is universal in *B. neritina*. *B. neritina* is abundant and readily accessible. There is a large base of knowledge about *B. neritina* biology: it is the *E. coli* of bryozoan biology. It can be maintained in the laboratory, and even raised from larvae through reproductive maturity in the laboratory. The mechanism of transmission of "*E. sertula*" is well established, and is not complicated by recruitment from free-living populations. The association makes a single identified type of natural product, the bryostatins. Bryostatins are complex polyketides, a group that contains numerous molecules with a wide range of bioactivities and applications. The biosynthesis of complex polyketides is well understood and the molecular biology of complex polyketide biosynthesis is a very active area of current research (Cane *et al.*, 1998). Although it would be more convenient if *"E. sertula"* were readily cultivable by conventional techniques, if it were, it would not be representative of many natural product symbioses. The numerous virtues of the *B. neritina*/*"E. sertula"* association means that testing the hypothesis that *"E. sertula"* is the biosynthetic source of bryostatins should be a high priority.

Sponges

In the early Cambrian Period, before the rise of scleractinian corals, sponges thrived as reef-builders living in close association with bacterial mats (De Freitas, 1991; Brunton and Dixon, 1994; Zhang and Pratt, 1994; Riding and Zhuravlev, 1995). These simple metazoans have maintained their prokaryotic ties into the modern world, filtering bacteria from seawater as food (Turon *et al.*, 1997) and harboring large and diverse microbial populations (Wilkinson, 1984). In many sponges, the bacteria have been shown to play a role in the lives of their host, either through processing of waste products, (Beer and Ilan, 1998) transfer of nutrients to their hosts, (Schumann-Kindel *et al.*, 1997) or production of secondary metabolites (Unson *et al.*, 1994). Thus, sponges are characterized by complex microbial communities that contrast strongly with the simple symbiotic systems of bryozoans. This complexity presents opportunities for interactions that could lead to greater diversity of natural products, but can also complicate microbiological analysis. Sponges have provided more natural products than any other phylum of marine invertebrate, and many of the natural products have potent bioactivities and unprecedented molecular architectures. A microbial origin has been proposed for a number of "sponge" metabolites, but only in a few instances have these hypotheses been tested experimentally. As is the case for bryostatins, many sponge metabolites are limited by their low availability from natural sources. Studies of sponge-microbe symbioses could lead to a better supply of potential pharmaceuticals, and in addition, there is much to learn from these ancient and diverse symbioses about the formation and maintenance of relationships between bacteria and metazoans.

Symbionts in Sponges
As the simplest of metazoans, sponges (Phylum: Porifera) have an extremely long evolutionary history, with a fossil record that indicates a Precambrian origin for the phylum (Brunton and Dixon, 1994). The history of sponges is tightly coupled to microorganisms, which they filter from seawater with high efficiency (75-99%)(Turon *et al.*, 1997). In addition to their role as a primary food source for poriferans, bacteria have also interacted with sponges to produce reef-like structures (De Freitas, 1991; Brunton and Dixon, 1994; Zhang and Pratt, 1994; Riding and Zhuravlev, 1995). Sponge-microbe associations in ancient reefs have been reviewed in detail (Brunton and Dixon, 1994).

The sponges that formed ancient reefs or are well documented in the Phanerozoic Era fossil record tend to have strongly reinforced skeletons. For that reason, the major sponge fossils are hexactinellids or from the demosponge order Lithistida, both of which have strong siliceous spicules and occur widely in the deep ocean today. Unfortunately, the order Lithistida is polyphyletic, with a taxonomy based only on a spicule type (desma), and thus does not reflect the true evolutionary relationship of these sponges (McInerney and Kelly-Borges, 1998). Because of the uncertain relationship between fossil and living sponges, it is difficult to determine the importance of the early sponge-microbe associations in the evolution of living sponges.

Despite the lack of a clear co-evolution of sponges and microbes in the fossil record, numerous associations between microbes and extant sponges have been described. Sponges feed on microbes, and also harbor large populations (up to 40% of body mass) of microorganisms, both intra- and extra-cellularly (Vacelet and Donadey, 1977). Bacterial density is particularly high in massive sponges with a large mesohyl, and cyanobacteria often coat the outer layers of shallow, exposed sponges. The literature on the ecological significance of microbial sponge symbionts through 1987 has been reviewed by Wilkinson (1987). Until that time, sponges were studied primarily for their relationship with photosynthetic organisms such as cyanobacteria and dinoflagellates, and nutrient transfer relationships were demonstrated in a number of cases. Nitrogen fixation and the presence of facultative and obligate anaerobes in marine sponges were also areas of active research. In some of these studies, it was shown that sponges derive a significant amount of their nutrients from microbial symbionts, although in one recent study the transfer of photosynthate from cyanobacteria to the sponge *Theonella swinhoei* was shown to be of minor importance (Beer and Ilan, 1998). Possibly, the most novel trophic transfer from microbe to sponge is a symbiosis between the deep-sea carnivorous sponge *Cladorhiza* sp. and a methane-oxidizing bacterium (Vacelet *et al.*, 1996). The sponges live near mud volcanoes that leach methane, and they seem to derive a significant amount of nutrition from their microbial symbionts. However, without cultivable bacteria or taxonomic data derived from molecular biology, the evolutionary significance of the symbiosis is uncertain.

During the 1970s and 1980s, investigators also studied the type and specificity of sponge-microbe associations, often using microscopy to show the presence of specific symbiont morphologies within specific sponges. The molecular basis of symbiosis was also probed, albeit to a lesser extent. Bacteria which live symbiotically with sponges, can be passed through their feeding chambers without being digested. This suggested some sort of encapsulation or recognition process (Wilkinson, 1987). In the demosponge *Halichondria panicea*, an association with the microbe *Pseudomonas insolita* may be lectin-based (Müller *et al.*, 1981). Wilkinson found an immunological basis for symbiosis in some sponges, which he claimed as evidence of a Precambrian origin for many symbioses (Wilkinson, 1984).

A major problem with the early studies on sponge-microbe symbiosis was that most microorganisms were uncultured or unculturable, so descriptions of symbioses usually relied either on morphology of symbionts or chemical measurements of nutrient transfer. Even in the cases where putative symbionts could be cultured, the ecological relevance of symbiosis could not be determined. The period following Wilkinson's review has been marked by the ascendance of molecular biological techniques in environmental microbiology, which have allowed investigators to focus on uncultured microorganisms.

The application of molecular biology to sponge-microbe symbiosis is yielding results that could not have been obtained by classical microbiological methods. The discovery of a member of the Archaea living specifically within a sponge similar to *Axinella mexicana* was a particularly exciting find (Preston *et al.*, 1997). The archaeal microorganism, *Cenarchaeum symbiosum* (P: Crenarchaeota), lives at a relatively cold 10 °C and is therefore considered psychrophilic. Subsequent *in situ* hybridization experiments showed which microorganism in the sponge was archaeal and allowed localization of the symbiont.

Another sequence-based study of demosponges aimed to elucidate the major microorganisms within the marine sponges *Chondrosia reniformis* and *Petrosia ficiformis* (Schumann-Kindel *et al.*, 1997). *In situ* hybridization probes specific to Archaea, the subclasses of Proteobacteria, Flavobacteria-Cytophaga, and sulfate-reducing bacteria was used to survey these sponges. Both sponges contained mainly γ-sublcass Proteobacteria, according to the authors, with sulfate reducing (δ-subclass) Proteobacteria also being present in significant numbers. γ-Proteobacteria were the only bacterial isolates obtained using aerobic enrichment media. Sulfate-reducing bacteria were postulated to play a role in the mineralization of dead sponge tissue.

In some cases, the picture of symbiosis that is emerging is markedly different from studies using cultured organisms. For instance, in the same species, *Halichondria panicea*, used by Müller *et al.* in lectin studies, sequence retrieval without cultivation led to microorganisms wholly different from the cultured *Pseudomonas* sp. described earlier (Althoff *et al.*, 1998). SSU rRNA sequences from all individuals of *H. panicea* collected led to the identification of strains of *Rhodobacter* (α-Proteobacteria) as the dominant species. A symbiotic relationship between the sponge and these microbes was suggested based on their ubiquitous occurrence in *H. panicea*. The activities of these bacteria and their effects on the host remain unknown.

Natural Products in Sponge-microbe Symbiosis

Natural products chemists have been interested in sponge-microbe symbioses because of the possibility that the diverse, bioactive chemical structures found in sponges might be produced by microorganisms (Faulkner *et al.*, 1993; Kobayashi and Ishibashi, 1993). More natural products have been reported from sponges than from any other marine invertebrate phylum, and many of the most promising pharmaceuticals and agents for cell biological research were isolated from sponges.

Unfortunately, despite the scientific and technological benefits of determining the source organisms of interesting and bioactive metabolites, a microbial origin of sponge compounds has rarely been demonstrated (Unson *et al.*, 1994; Bewley *et al.*, 1996). Most of the literature is purely speculative, based on similarities, however slight, between compounds from sponges and those from cultivated microorganisms, especially cyanobacteria

(Kobayashi and Ishibashi, 1993). Several researchers have attempted to culture microorganisms from invertebrates in the hopes of obtaining some of these bioactive compounds (Pietra, 1997). Although they have been successful in the discovery of novel natural products, this research has rarely demonstrated the presence of sponge metabolites in the microbial isolates. In one case the same compound was found in a *Hyatella* sp. sponge as in a *Vibrio* sp. cultured from that sponge (Oclarit *et al.*, 1998). These results demonstrate that traditional culturing approaches are not generally applicable to the environmental problems of sponge-microbe symbiosis. Two cases of symbiont production of sponge compounds have been clearly established. Both studies relied on cell fixation and physical separation techniques, bypassing the problem of culturing symbiotic microorganisms. Unson separated cyanobacterial symbionts from the sponge *Dysidea herbacea* by flow cytometry and showed that chlorinated amino acid derivatives could only be found in the cyanobacterial fraction, while terpenes were localized in the sponge cell fraction (Unson *et al.*, 1994). The symbiont was typed as *Oscillatoria spongeliae* based on morphological characteristics. In another study, Bewley determined the cellular locations of two major bioactive metabolites, theopalauamide and swinholide A, from the lithistid sponge *Theonella swinhoei* (Bewley *et al.*, 1996) (Figure 2). Contrary to the expectation that the metabolites would be produced by cyanobacteria, it was found that the modified peptide theopalauamide was found only in filamentous microorganisms without photosynthetic pigments, while the polyketide swinholide A was located solely in the unicellular bacterial fraction. *T. swinhoei* is a massive sponge that is packed with bacteria of many different species, so the exact microorganism producing swinholide A could not be determined, but in TEM pictures the filamentous microorganism was thought to resemble *Beggiatoa* sp., which are γ-Proteobacterial sulfide oxidizers. Recently, the theopalauamide-containing filaments were shown, by 16S rRNA sequencing and *in situ* hybridization, to be δ-proteobacteria related to the myxobacteria. The filaments were given the provisional taxonomic status "*Candidatus* Entotheonella palauensis". This organism can be propagated outside of the sponge host in a mixed bacterial culture (Schmidt *et al.*, in press).

Despite a good deal of speculation on the microbial origin of sponge metabolites, relatively little has been conclusively demonstrated. The number of cell separation studies on sponges is increasing, but a complete understanding of the source organisms of sponge compounds is still a distant goal. Molecular biological techniques, specifically the location of biosynthetic genes within certain cell types, are just beginning to be applied to the source question. However, no studies have addressed the evolution and ecological importance of sponge-microbe symbioses in the production of natural products. An important initial step in investigating these symbioses is to address the identity of chemically important microbes in sponges.

Lithistid Sponges
In comparison to bryozoans, much more is known about the biology and natural products chemistry of sponges. This presents great opportunities for symbiosis research, but also potentially complicates the search for a model

group of sponges from the large number of described taxa. One such model group can be found in the polyphyletic order Lithistida. In this taxon, natural products may be used, in concert with biological tools, to probe the evolution of bacterial-poriferan symbiosis. The order contains a large number of diverse, bioactive molecules that have been ascribed microbial origins, and in one case a bacterial source has been proven (Bewley *et al.*, 1996). Because it is relatively easy to determine the presence of known compounds in a sponge, the distribution and specificity of metabolites in this order can be mapped. By comparing this data with sponge taxonomy and molecular phylogeny of sponge symbionts, hypotheses of the nature of these symbioses can be advanced.

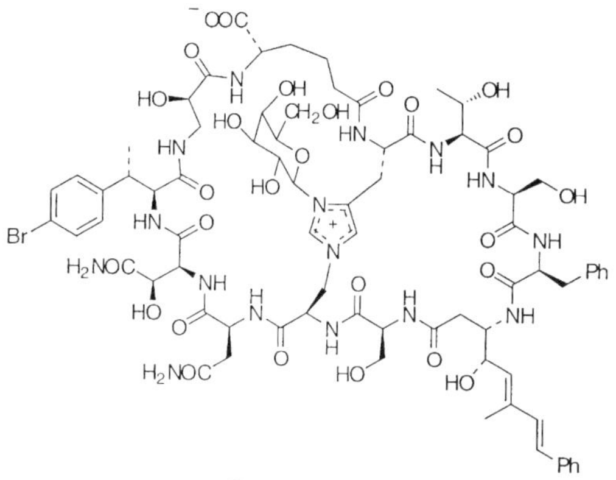

Theopalauamide

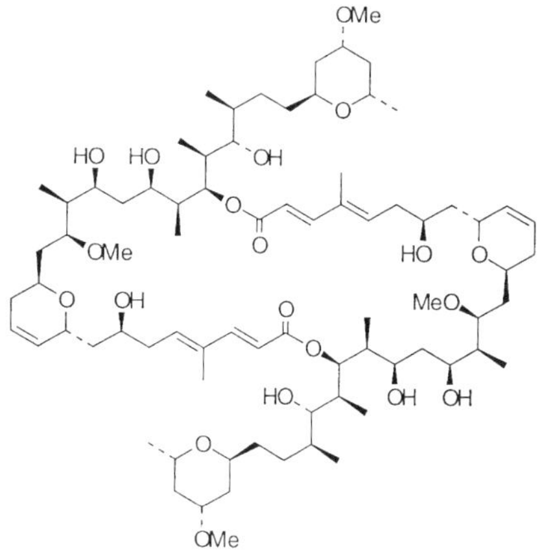

Swinholide A

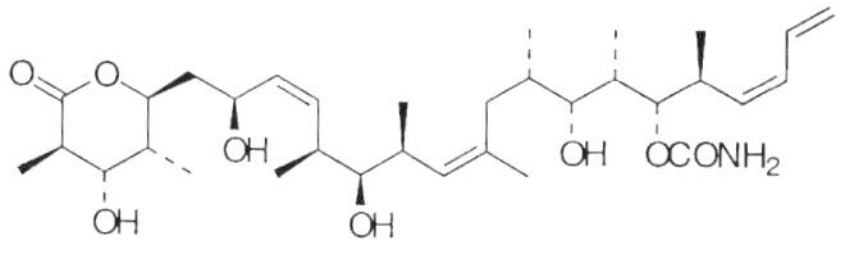

Discodermolide

Figure 2. Sponge Natural Products

In cases in which a link between a metabolite and specific bacteria has been proven, natural products can provide a convenient signal of bacterial-sponge symbiosis. Compounds from the order Lithistida are particularly good markers, since they are usually present in relatively large amounts, often have unusual bioactivities, and can be relatively easily purified or identified in crude extracts. Many lithistid compounds are probably of microbial origin (Fusetani and Matsunaga, 1993), and in the case of *T. swinhoei* there is strong evidence supporting bacterial production (Bewley *et al.*, 1996). In some cases, production of certain compounds seems species-specific, while other compounds are found in a number of unrelated lithistids, or even in other sponge orders. This opens the possibility of a large number of types, ages, and plasticities of symbiotic interactions. Physical or biochemical factors common to unrelated lithistid sponges, rather than a species-specific factor, may promote the growth of certain types of bacteria. By attempting to construct gene probes for the bacteria that produce lithistid compounds, symbiosis in this group of sponges can be characterized and also provide a model for other sponge groups.

From the biotechnological perspective, lithistids also make good subjects for symbiosis research because they often contain highly bioactive compounds with pharmaceutical or research potential. Supreme among lithistid metabolites is the polyketide discodermolide from the sponge *Discodermia dissoluta* (Figure 2). The anti-tumor activity of discodermolide has been compared to that of taxol. Polyketide metabolites swinholide A from *T. swinhoei* and calyculin A from *Discodermia calyx* are valuable agents for research in cell biology because of their potent and specific activities. Swinholide A stabilizes tubulin dimers, while calyculin A is a potent protein phosphatase II inhibitor. Modified peptides, most of which are probably produced non-ribosomally by peptide synthetases, are also known for their potent bioactivities. The antitumor cyclotheonamides, from *Theonella* sp., exemplify these compounds. The status of chemical research in the order Lithistida as of early 1997 has been reviewed (Bewley and Faulkner, 1998).

Lithistid compounds often share characteristics with microbial metabolites, particularly with those of the cyanobacteria (Fusetani and Matsunaga, 1993; Bewley and Faulkner, 1998). Once Bewley *et al.* (1996) showed that the major metabolites of *Theonella swinhoei*, a lithistid sponge, were localized in symbiotic bacteria, the hypothesis that many compounds in the order are microbe-derived gained some experimental weight. Although it would require a massive series of cell-separation experiments on a large number of lithistid sponges to test the hypothesis, it is nonetheless a reasonable assumption when choosing targets for molecular biological research. It must be noted that the authors only showed the existence of the compounds in the bacteria and not bacterial production, but it is reasonable to presume that these large metabolites are not completely transferred from one cell type to another.

Bewley hypothesized that lithistid peptides containing ω-phenyl-β-amino acids were produced by filamentous microorganisms, similar to *"E. palauensis"*, the putative producer of theopalauamide in *T. swinhoei*. This hypothesis was based on her observation of the correlation between the two properties in a large number of lithistid sponges (Bewley, 1995). Thus, gross morphology of the symbiont as well as chemistry could be used in the analysis of theopalauamide origin. However, peptide-producing sponges do not always contain filamentous microorganisms. Samples of *Aciculites* sp. containing compounds in the aciculitin family do not contain filamentous microorganisms, and representatives of *Microscleroderma* sp. containing microsclerodermins sometimes lack filamentous microorganisms. Recently, PCR primers based on the 16S rRNA sequence of *"E. palauensis"*, the theopalauamide-containing δ-proteobacterium in *T. swinhoei*, were applied to several theonellid sponges. From some samples of *T. swinhoei*, the primers amplified sequences that are very similar to the *"E. palauensis"* sequence. These sponges contained the peptides theonegramide and theonellamides, which have structures similar to the *"E. palauensis"* metabolite, theopalauamide. In contrast, theonellids containing the unrelated aurantosides and mozamides yielded somewhat more distantly related sequences (Schmidt *et al.*, in press). It would be interesting to determine whether the aurantosides and mozamides are localized in bacteria, as suggested by the sequence data. A definitive study of the phylogenetic relationships among theonellid sponges, their symbionts and their chemistry would be very illuminating.

The origin of other metabolites (mainly polyketides) has been more complicated to sort out. The prime example in this group of compounds is swinholide A, which was localized by Bewley *et al.* (1996) in unicellular bacterial fractions of *T. swinhoei* containing a mixture of bacteria. Thus it was impossible to identify a specific bacterial source. In addition, the compound and a related chemical were subsequently found in two sponges that are unrelated: *Ircinia* sp. and *Tedania diversiraphidiphora* from the order Dictyoceratida and *Lamellomorpha strongylata* from the order Epipolasida (Dumdei *et al.*, 1997). In the absence of lateral transfer of biosynthetic genes or in the unlikely event that the complex structure could be reached by convergent evolution, these results indicate that the symbiosis leading to the production of swinholide is not species- or even order-specific. Unfortunately, it is difficult to track down which bacterial species is actually producing swinholide because of the large number of unicellular bacteria present in these filter-feeders. This illustrates the problems inherent in investigating the complex microbial communities found in sponges. Another instance of polyketide metabolites found in different orders occurs with the structurally similar calyculins (Kato *et al.*, 1986) and clavosines (Fu *et al.*, 1998) from *Discodermia calyx* (O: Lithistida) and *Myriastra clavosa* (O: Choristida), respectively. These similar compounds and related metabolites are found in apparently unrelated sponges. It would be interesting to determine if these and other similar compounds are always produced by similar bacteria, or if convergent evolution or lateral gene transfer are possibilities. Studies of these hypotheses become accessible through the combination of chemical and biological data. Ultimately, such studies could lead to a better understanding of the evolutionary importance of these compounds and to better culturing methods for non-specific symbionts, by comparison of common factors in the host enviroment.

The polyphyletic nature of the order Lithistida may confound early attempts at rationalizing the evolution of symbiosis. A clear picture of sponge phylogeny is necessary to fully investigate chemical and symbiont specificity, and elucidate the evolution of natural product symbioses. Some preliminary results of phylogenetic analysis of listhistids are available (McInerney and Kelly-Borges, 1998). For instance, sponges from the genera *Plakinalopha* and *Theonella*, formerly thought not to be closely related, are actually closest relatives: some *Theonella* spp. are more closely related to certain *Plakinalopha* spp. than to other theonellids. This clarifies chemical observations that some members of both *Theonella* and *Plakinolopha* contain aurantosides and mozamides or related peptides as their major bioactive metabolites. The compound classes are very similar to compounds produced by streptomycetes and cyanobacteria, respectively. It would be interesting to determine whether or not the compounds are actually produced by bacteria and whether such a symbiosis holds in members of both sponge genera.

The listhistid sponges have inherent disadvantages: complex and variable microbial populations, difficulty in maintaining them in captivity, and lack of knowledge about symbiont transmission. Nonetheless, they have important advantages as a model system complementary to a more tractable model such as *B. neritina*. The advantages of sponges include proof that microorganisms are the biosynthetic source of two natural products, and high levels of natural products, facilitating chemical analysis. In addition, studies on the genetics of biosynthesis of swinholide A, a polyketide, and theopalauamide, a non-ribosomal peptide, will benefit from the substantial body of research available on the biosynthesis of these classes of compounds. Finally, sponges are a rich source of compounds, a number of which may be made by microorganisms. Despite their microbiological complexity, it is imperative to develop appropriate models for study of these symbioses.

Research Opportunities

Surveying and identifying microbial symbionts present in host invertebrates is fundamentally important in cases in which biosynthesis of a natural product by a symbiont is hypothesized. Techniques for molecular survey are now standard in microbial ecology, and can be applied to symbiotic systems. Ribosomal RNA genes are the most frequently used because of the large databases available and the inherent phylogenetic utility of these genes. Amplification of rRNA genes followed by cloning or denaturing gradient gel electrophoresis can be used to evaluate the relative abundance of particular bacteria, and comparison between samples can distinguish consistently associated organisms from transient colonizers. Sequences of the genes amplified reveals the phylogenetic groups to which the organisms belong and can provide specific probes for *in situ* hybridization, allowing sequences to be linked to specific cells. This approach has been used successfully in *B. neritina* and *T. swinhoei*, and has great potential for improving our understanding of sponge symbioses.

Complementary approaches are required for evaluating the biosynthetic source of natural products. Cell dissociation and chemical analysis can be very powerful, as shown by the discovery of sponge natural products in bacterial cells (Unson *et al.*, 1994; Bewley *et al.*, 1996). In cases in which gross dissociation is not feasible, it will be important to develop *in situ* techniques for localizing specific natural products to bacterial or host cells, or co-localizing with particular bacterial populations using specific rRNA probes. Antibodies might be used for this purpose, or chemical techniques might be developed. An important caveat is that highly mobile compounds or those that are actively transported within the host could result in no strong localization and make conclusions about the biosynthetic source difficult.

Natural products chemistry investigations assist in formation of hypotheses about sponge-microbe symbioses. By positively correlating certain metabolites and metabolite classes with microbial symbionts, the distribution of metabolites can provide clues about the nature of the symbiosis. For instance, swinholide A has been isolated from a mixed bacterial fraction in *T. swinhoei*, and related compounds are found in a number of unrelated sponges. Therefore, it could be proposed that the symbiont prefers a sponge host, but it is probably present in seawater and can be transmitted between sponges of various species. Since sponges are efficient filter feeders, any bacterium that can resist the sponge digestive process and immune response can successfully colonize a variety of sponges. Using a combination of molecular biological and chemical techniques, the identity of this microorganism could be determined, and isolation of swinholide from unrelated sponges could serve as a starting point for detailed studies on the symbiosis. In another example from *T. swinhoei*, theopalauamide and related compounds, and *"E. palauensis"*-like bacterial 16S rRNA sequences, have only been found in theonellid sponges. Therefore, it could be proposed that the symbiosis is probably more species-specific, involving long-term co-evolution. In depth investigation of evolution of invertebrate symbioses and natural product biosynthesis will require careful phylogenetic analysis of both symbiont and host genes combined with chemical studies.

Microbial ecology in general, as well as this area of research, will benefit greatly from the development of novel methods of cultivation of symbionts and other environmental organisms. The hallmark of the bacteria that are most commonly cultivated and studied in the laboratory is their great adaptability to different conditions. We do not know the reasons why the majority of symbionts (such as *"E. sertula"* in *B. neritina*), and environmentally dominant bacteria are difficult to cultivate, but requirements for highly specific conditions are likely to be part of the problem. We need to develop effective methods to bring these organisms into, if not traditional pure cultures, at least stable means of propagation that allow laboratory study. Symbioses are excellent model systems for developing such approaches because of stable, specific environmental conditions provided by the host.

Even in cases in which natural product localization is not feasible and the bacteria are resistant to cultivation, elucidation of the biosynthetic pathway through biochemical approaches may be possible and can reveal a likely symbiotic origin through comparative sequence analysis. Cloning of biosynthetic genes, based on inferences about biosynthesis based on chemical structure, is likely to be more straightforward for bacterial symbionts than host invertebrates, because of the way bacterial genes are organized. This may lead to a satisfactory solution to the problem of supply without cultivating the organism. Logical targets for isolation of biosynthetic genes include polyketide synthases for bryostatins in *B. neritina* and swinholides in lithistids, and non-ribosomal peptide synthases for theopalauamide-like peptides in lithistids. Research on the regulation of biosynthesis will reveal the intimate details of the symbiotic relationship and can be exploited for improving yield. Finally, when biosynthetic genes from symbionts become available, this will facilitate the creation of novel compounds with potentially valuable bioactivities by manipulating and combining genes from symbionts and other organisms.

Conclusion

Researchers have just begun to investigate the role of symbionts in the biosynthesis of natural products in marine invertebrates. The potential for molecular microbiology to produce fundamental biological discoveries is tantalizing, and the possibilities for enhancing the availability of these otherwise inaccessible compounds would be a major accomplishment for microbial biotechnology. The study of invertebrate natural product symbioses is poised to emerge as an exciting new field for molecular microbiology.

Acknowledgements

Support for this work was provided by the National and California Sea Grant College Programs under grant number NA36RG0537 project number RMP-61, and grant number NA66RG0477, project number R/MP-84A (MH), and the National Science Foundation (JF).

References

Althoff, K., Schütt, C., Steffen, R., Batel, R., and Müller, W.E.G. 1998. Evidence for a symbiosis between bacteria of the genus *Rhodobacter* and the marine sponge *Halichondria panicea*: Harbor also for putatively-toxic bacteria? Mar. Biol. 130: 529-536.

Anthoni, U., Nielsen, P.H., Perieira, M., and Christophersen, C. 1990. Bryozoan secondary metabolites: a chemotaxonomical challenge. Comp. Biochem. Physiol. 96B: 431-437.

Bai, R., Cichacz, Z.A., Herald, C.L., Pettit, G.R., and Hamel, E. 1993. Spongistatin 1, a highly cytotoxic, sponge-derived, marine natural product that inhibits mitosis, microtubule assembly, and the binding of vinblastine to tubulin. Mol. Pharmacol. 44: 757-66.

Baldwin, N.G., Rice, C.D., Tuttle, T.M., Bear, H.D., Hirsch, J.I., and Merchant, R.E. 1997. *Ex vivo* expansion of tumor-draining lymph node cells using compounds which activate intracellular signal transduction. I. Characterization and *in vivo* anti-tumor activity of glioma-sensitized lymphocytes. J. Neurooncol. 32: 19-28.

Beer, S., and Ilan, M. 1998. *In situ* measurements of photosynthetic irradiance responses of two Red Sea sponges growing under dim light conditions. Mar. Biol. 131: 613-617.

Bewley, C.A. 1995. New antifungal and cytotoxic cyclic peptides and studies of the bacterial symbionts of lithistid sponges. Thesis. Scripps Institution of

Oceanography, University of California, San Diego, La Jolla.

Bewley, C.A., and Faulkner, D.J. 1998. Lithistid sponges: Star performers or hosts to the stars? Angew. Chem. Int. Ed. Engl. 37: 2162-2178.

Bewley, C.A., Holland, N.A., and Faulkner, D.J. 1996. Two classes of metabolites from *Theonella swinhoei* are localized in distinct populations of bacterial symbionts. Experientia. 52: 716-722.

Boyle, P.J., Maki, J.S., and Mitchell, R. 1987. Mollicute identified in novel association with aquatic invertebrate. Current Microbiol. 15: 85-89.

Brunton, F.R., and Dixon O.A. 1994. Siliceous sponge-microbe biotic associations and their recurrence through the Phanerozoic as reef mound constructors. Palaios. 9: 370-387.

Buddemeier, R.W. 1994. Symbiosis, calcification, and environmental interactions. Bulletin de l'Institut Oceanographique (Monaco). p. 119-135.

Cane, D.E., Walsh, C.T., and Khosla, C. 1998. Harnessing the biosynthetic code: Combinations, permutations, and mutations. Science. 282: 63-68.

Caponigro, F., French, R.C., and Kaye, S.B. 1997. Protein kinase C: a worthwhile target for anticancer drugs? Anticancer Drugs. 8: 26-33.

Cavanaugh, C.M. 1994. Microbial symbiosis: Patterns of diversity in the marine environment. American Zoologist. 34: 79-89.

Christophersen, C. 1985. Secondary metabolites from bryozoans. Acta. Chem. Scand. 39: 517-529.

Correale, P., Caraglia, M., Fabbrocini, A., Guarrasi, R., Pepe, S., Patella, V., Marone, G., Pinto, A., Bianco, A.R., and Tagliaferri, P. 1995. Bryostatin 1 enhances lymphokine activated killer sensitivity and modulates the beta 1 integrin profile of cultured human tumor cells. Anticancer Drugs. 6: 285-90.

Dahlberg, C., Bergström, M., and Hermansson, M. 1998. *In situ* detection of high levels of horizontal plasmid transfer in marine bacterial communities. Appl. Environ. Microbiol. 64: 2670-2675.

Davidson, S.K., and Haygood, M.G., 1999. Identification of sibling species of the bryozoan *Bugula neritina* that produce different anticancer bryostatins and harbor distinct strains of the bacterial symbiont "*Candidatus* Endobugula sertula". Biol. Bull. In press.

De Freitas, T.A. 1991. Ludlow (Silurian) lithistid and hexactinellid sponges, Cape Phillips Formation, Canadian Arctic. Can. J. Earth. Sci. 28: 2042-2061.

Dumdei, E.H., Blunt, J.W., Munro, M.H.G., and Pannell, L.K. 1997. Isolation of calyculins, calyculinamides, and swinholide H from the New Zealand deep-water marine sponge *Lamellomorpha strongylata*. J. Org. Chem. 62: 2636-2639.

Faulkner, D.J., He, H.Y., Unson, M.D., and Bewley, C.A. 1993. New metabolites from marine sponges: Are symbionts important? Gazz. Chim. Ital. 123: 301-307.

Fleming, M.D., Bear, H.D., Lipshy, K., Kostuchenko, P.J., Portocarero, D., McFadden, A.W., and Barrett, S.K. 1995. Adoptive transfer of bryostatin-activated tumor-sensitized lymphocytes prevents or destroys tumor metastases without expansion in vitro. J. Immunother. Emphasis Tumor Immunol. 18: 147-55.

Fu, X., Schmitz, F.J., Kelly-Borges, M., McCready, T.L., and Holmes, C.F.B. 1998. Clavosines A-C from the marine sponge *Myriastra clavosa*: Potent cytotoxins and inhibitors of protein phosphatases 1 and 2A. J. Org. Chem. 63: 7957-7963.

Fusetani, N., and Matsunaga, S. 1993. Bioactive sponge peptides. Chem. Rev. 93: 1793-1806.

Grant, S., Traylor, R., Pettit, G.R., and Lin, P.S. 1994. The macrocyclic lactone protein kinase C activator, bryostatin 1, either alone, or in conjunction with recombinant murine granulocyte-macrophage colony-stimulating factor, protects Balb/c and C3H/HeN mice from the lethal *in vivo* effects of ionizing radiation. Blood. 83: 663-667.

Haygood, M.G. 1993. Light organ symbioses in fishes. Crit. Rev. Microbiol. 19: 191-216.

Haygood, M.G., and Davidson S.K. 1997. Small subunit ribosomal RNA genes and *in situ* hybridization of the bacterial symbionts in the larvae of the bryozoan *Bugula neritina* and proposal of "*Candidatus* Endobugula sertula". Appl. Env. Microbiol. 63: 4612-4616.

Higa, T., Tanaka, J.-I., and Komesu, M. 1992. Miyakolide: A bryostatin-like macrolide from a sponge, *Polyfibrospongia* sp. J. Am. Chem. Soc. 114: 7587-7588.

Hirata, Y., and Uemura, D. 1986. Halichondrins - antitumor polyether macrolides from a marine sponge. Pure Appl. Chem. 58: 701-710.

Hopwood, D.A. 1997. Genetic contributions to understanding polyketide synthases. Chem. Rev. 97: 2465-2497.

Hugenholtz, P., Goebel, B.M., and Pace, N.R. 1998. Impact of culture-independent studies on the emerging phylogenetic view of bacterial diversity. J. Bacteriol. 180: 4765-4774.

Kageyama, M., Tamura, T., Nantz, M.H., Roberts, J.C., Somfai, P., Whritenour, D.C., and Masamune, S. 1990. Synthesis of bryostatin-7. J. Am. Chem. Soc. 112: 7407-7408.

Kato, Y., Fusetani, N., Matsunaga, S., and Hashimoto, K. 1986. Calyculin A, a novel antitumor metabolite from the marine sponge *Discodermia calyx*. J. Am. Chem. Soc. 108: 2780-2781.

Katz, L., and Donadio S. 1993. Polyketide synthesis: prospects for hybrid antibiotics. Annual Rev. Microbiol. 47: 875-912.

Keough, M.J. 1989. Dispersal of the bryozoan *Bugula neritina* and effects of adults on newly metamorphosed juveniles. Marine Ecol. Prog. Series. 57: 163-171.

Kerr, R.G., Lawry J., and Gush, K.A. 1996. *In vitro* biosynthetic studies of the bryostatins, anti-cancer agents from the marine bryozoan *Bugula neritina*. Tetrahedron Lett. 37: 8305-8308.

Kobayashi, J., and Ishibashi M. 1993. Bioactive metabolites of symbiotic marine microorganisms. Chem. Rev. 93: 1753-1769.

Kobayashi, M., Tanaka, J., Katori, T., and Kitagawa, I. 1990. Marine natural products. XXIII. Three new cytotoxic dimeric macrolides, swinholides B and C and isoswinholide A, congeners of swinholide A, from the Okinawan marine sponge *Theonella swinhoei*. Chem. Pharm. Bulletin. 38: 2960-6.

Kraft, A.S. 1993. Bryostatin 1: will the oceans provide a cancer cure? J. Natl. Cancer Inst. 85: 1790-2.

Kraft, A.S., Woodley, S., Pettit, G.R., Gao, F., Coll, J.C., and Wagner, F. 1996. Comparison of the antitumor activity of bryostatins 1, 5, and 8. Cancer

Chemother. Pharmacol. 37: 271-8.

Lind, D.S., Tuttle, T.M., Bethke, K.P., Frank, J.L., McCrady, C.W., and Bear, H.D. 1993. Expansion and tumour specific cytokine secretion of bryostatin-activated T-cells from cryopreserved axillary lymph nodes of breast cancer patients. Surg. Oncol. 2: 273-82.

Lindquist, N., and Hay, M.E., 1996. Palatability and chemical defense of marine invertebrate larvae. Ecological Monographs. 66: 431-450.

Lipshy, K.A., Kostuchenko, P.J., Hamad, G.G., Bland, C.E., Barrett, S.K., and Bear, H.D. 1997. Sensitizing T-lymphocytes for adoptive immunotherapy by vaccination with wild-type or cytokine gene-transduced melanoma. Ann. Surg. Oncol. 4: 334-41.

Litaudon, M., Hart, J.B., Blunt, J.W., Lake, R.J., and Munro, M.G.H. 1994. Isohomohalichondrin B, a new antitumor polyether macrolide from the New Zealand deep-water sponge Lissodendoryx sp. Tetrahedron Lett. 35: 9435-9438.

Lutaud, G. 1964. Sur le structure et le rôle des glandes vestibulaire et sur la nature de certain organ de la cavité cystidienne chez le bryozoaires chilostomes. Cahiers de Biologie Marine. 5: 201-231.

Lutaud, G. 1965. Sur la présence de microorganismes spécifiques dans le glandes vestibulaires et dans l'aviculaire de Palmicellaria skenei (Ellis et Solander). Bryozoaires chilostome. Cahiers de Biologie Marine. 6: 181-190.

Lutaud, G. 1969. La nature des corps funiculaires des cellularines, bryozoaires chilostomes. Arch. Zool. Exp. Gen. 110: 5-30.

Lutaud, G. 1986. L'infestation du myoépithelium de l'esophage par des microorganismes pigmentés et la structure des organes à bactéries du vestibule chez le Bryozoaire Chilostome Palmicellaria skenei (E. et S.). Can. J. Zool. 64: 1842-1851.

McFall-Ngai, M.J. 1994. Animal-bacterial interactions in the early life history of marine invertebrates: The Euprymna scolopes-Vibrio fischeri symbiosis. Am. Zool. 34: 554-561.

McInerney, J., and Kelly-Borges M. 1998. Phylogeny of lithistid sponges. Origin and Outlook: 5th Internat. Sponge Symp., Brisbane, Australia.

Müller, W.E.G., Zahn, R.K., Kurelec, B., Lucu, C., Müller, I., and Uhlenbruck, G. 1981. Lectin, a possible basis for symbiosis between bacteria and sponges. J. Bacteriol. 145: 548-558.

Murray, R.G.E., and Stackebrandt, E. 1995. Taxonomic Note: Implementation of the provisional status Candidatus for incompletely described procaryotes. Internat. J. Syst. Bacteriol. 45: 186-187.

Nowak-Thompson, B., Gould, S.J., and Loper, J.E. 1997. Identification and sequence analysis of the genes encoding a polyketide synthase required for pyoluteorin biosynthesis in Pseudomonas fluorescens Pf-5. Gene. 204: 17-24.

Oclarit, J.M., Tamaoka, Y. Kamimura, K., Ohtan, S., and Ikegami, S. 1998. Andrimid, an antimicrobial substance in the marine sponge Hyatella produced by an associated Vibrio bacterium. In: Sponge Sciences: Multidisciplinary Perspectives. Y. Watanabe and N. Fusetani, eds. Springer-Verlag, Tokyo. p. 391-398.

Ojika, M., Yoshino, G., and Sakagami, Y. 1997. Novel ceramide 1-sulfates, potent DNA topoisomerase I inhibitors isolated from the Bryozoa *Watersipora cucullata*. Tetrahedron Lett. 38: 4235-4238.

Pettit, G.R. 1991. The bryostatins. Prog. Chem. Org. Nat. Prod. 57: 153-195.

Pettit, G.R., Cichaz, Z..A., Gao, F., Herald, C., Boyd, M.R., Schmidt, J.M., and Hooper, J.N.A. 1993. Isolation and structure of Spongistatin 1. J. Org. Chem. 58: 1302-1304.

Pettit, G.R., Gao, F., Blumberg, P.M., Herald, C.L., Coll, J.C., Kamano, Y., Lewin, N.E., Schmidt, M., and Chapuis, J.C. 1996. Antineoplastic agents. 340. Isolation and structural elucidation of bryostatins 16-18. J. Nat. Products. 59: 286-9.

Pettit, G.R., Gao, F., Sengupta, D., Coll, J.C., Herald, C.L., Doubek, D.L., Schmidt, J.M., Van Camp, J.R., Rudloe, J.J., and Nieman, R.A. 1991. Isolation and structure of bryostatins 14 and 15. Tetrahedron Lett. 22: 3601-3610.

Pettit, G.R., Herald, C.L., Doubek, D.L., and Herald, D.L. 1982. Isolation and structure of bryostatin 1. J. Am. Chem. Soc. 104: 6846-6847.

Pietra, F. 1997. Secondary metabolites from marine microorganisms. Bacteria, protozoa, algae, and fungi. Achievements and perspectives. Nat. Prod. Reports. 14: 453-464.

Pluda, J.M., Cheson, B.D., and Phillips, P.H. 1996. Clinical trials referral resource. Clinical trials using bryostatin-1. Oncology. 10: 740-2.

Preston, C.M., Wu, K.Y., Molinski, T.F., and DeLong, E.F. 1997. A psychrophilic crenarchaeon in habits a marine sponge: *Cenarchaeum symbiosum*, gen. nov., sp. nov. Proc. Natl. Acad. Sci. USA. 93: 6241-6246.

Reed, C.G., and Woollacott, R.M. 1983. Mechanisms of rapid morphogenetic movements in the metamorphosis of the bryozoan *Bugula neritina* (Cheilostomata, Cellularoidea): II. The role of dynamic assemblages in the pallial epithelium. J. Morphol. 177: 127-143.

Riding, R., and Zhuravlev A.Y. 1995. Structure and diversity of oldest sponge-microbe reefs: Lower Cambrian, Aldan River, Siberia. Geol. 23: 649-652.

Rowan, R. 1998. Diversity and ecology of zooxanthellae on coral reefs. J. Phycol. 34: 407-417.

Schaufelberger, D.E., Koleck, M.P., Beutler, J.A., Vatakis, A.M., Alvarado, A.B., Andrews, P., Marzo, L.V., Muschik, G.M., Roach, J., Ross, J.T., Lebherz, W.B., Reeves, M.P., Eberwein, R.M., Rodgers, L L., Testerman, R.P., Snader K.M., and Forenza, S. 1991. The large-scale isolation of bryostatin 1 from *Bugula neritina* following current good manufacturing practices. J. Nat. Products. 54: 1265-1270.

Scheid, C., Prendiville, J., Jayson, G., Crowther, D., Fox, B., Pettit, G.R., and Stern, P.L. 1994. Immunomodulation in patients receiving intravenous Bryostatin 1 in a phase I clinical study: comparison with effects of Bryostatin 1 on lymphocyte function *in vitro*. Cancer Immunol. Immunother. 39: 223-30.

Schmidt, E.W., Obraztsova, A.Y., Davidson, S.K., Faulkner, D.J., and Haygood, M.G. Identification of the antifungal peptide-containing symbiont of the marine sponge *Theonella swinhoei* as a novel δ-proteobacterium, "*Candidatus* Entotheonella palauensis". Mar. Biol. In press.

Schumann-Kindel, G., Bergbauer, M., Manz, W., Szewzyk, U., and Reitner,

J. 1997. Aerobic and anaerobic microorganisms in modern sponges: a possible relationship to fossilization-processes. Facies. 36: 268-272.

Schupp, T., Toupet, C., Cluzel, B., Neff, S., Hill, S., Beck, J.J., and Ligon, J.M. 1995. A *Sorangium cellulosum* (Myxobacterium) gene cluster for the biosynthesis of the macrolide antibiotic Soraphen A: cloning, characterization, and homology to polyketide synthase genes from actinomycetes. J. Bacteriol. 177: 3673-3679.

Shindo, T., Sato, A., Ksanuki, N., Hasegawa, K., Sato, S., Iwata, T., and Hata, T. 1993. 6H-anthra[1,9-bc]thiophene derivatives from a bryozoan, *Dakaria suboviodea*. Experientia. 49: 177-178.

Steube, K.G., and Drexler, H.G. 1993. Differentiation and growth modulation of myeloid leukemia cells by the protein kinase C activating agent bryostatin-1. Leuk. Lymphoma. 9: 141-8.

Sung, S.J., Lin, P.S., Schmidt-Ullrich, R., Hall, C.E., Walters, J.A., McCrady, C., and Grant, S. 1994. Effects of the protein kinase C stimulant bryostatin 1 on the proliferation and colony formation of irradiated human T-lymphocytes. Int. J. Radiat. Biol. 66: 775-83.

Taylor, L.S., Cox, G.W., Melillo, G., Bosco, M.G., and Espinoza-Delgado, I. 1997. Bryostatin-1 and IFN-gamma synergize for the expression of the inducible nitric oxide synthase gene and for nitric oxide production in murine macrophages. Cancer Res. 57: 2468-73.

Turon, X., Galera, J., and Uriz, M.J. 1997. Clearance rates and aquiferous systems in two sponges with contrasting life-history strategies. J. Exp. Zool. 278: 22-36.

Uemura, D., Takahashi, K., Yamamoto, T., Katayama, C., Tanaka, J., Okumura, Y., and Hirata, Y. 1985. Norhalichnodrin A; An antitumor polyether macrolide from a marine sponge. J. Am. Chem. Soc. 107: 4796-4798.

Unson, M.D., Holland, N.D., and Faulkner, D.J. 1994. A brominated secondary metabolite synthesized by the cyanobacterial symbiont of a marine sponge and accumulation of the crystalline metabolite in the sponge tissue. Mar. Biol. 119: 1-11.

Vacelet, J., and Donadey, C. 1977. Electron microscope study of the association between some sponges and bacteria. J. Exp. Mar. Biol. Ecol. 30: 301-314.

Vacelet, J., Fiala-Médioni, A. Fisher, C.R., and Boury-Esnault, N. 1996. Symbiosis between methane-oxidizing bacteria and a deep-sea carnivorous cladorhizid sponge. Mar. Ecol. Prog. Ser. 145: 77-85.

Wender, P.A., Debrabander, J., Harran, P.G., Jimenez, J.M., Koehler, M.F.T., Lippa, B., Park, C.M., Siedenbiedel, C., and Pettit, G.R. 1998. The design, computer modeling, solution structure, and biological evaluation of synthetic analogs of bryostatin 1. Proc. Natl. Acad. Sci. USA. 95: 6624-6629.

Wilkinson, C.R. 1984. Immunological evidence for the Precambrian origin of bacterial symbioses in marine sponges. Proc. R. Soc. Lond. 220: 509-517.

Wilkinson, C.R. 1987. Significance of microbial symbionts in sponge evolution and ecology. Symbiosis. 4: 135-146.

Woollacott, R.M. 1971. Attachment and metamorphosis of the Cheilo-ctenostome bryozoan *Bugula neritina* (Linne). J. Morph. 134: 351-382.

Woollacott, R.M. 1981. Association of bacteria with bryozoans larvae. Mar.

Biol. 65: 155-158.

Woollacott, R.M., and Zimmer R.L. 1975. A simplified placenta-like system for the transport of extraembryonic nutrients during embryogenesis of *Bugula neritina* (Bryozoa). J. Morph. 147: 355-378.

Woollacott, R.M., and Zimmer R.L. 1978. Metamorphosis of cellularioid bryozoans. In: Settlement and metamorphosis of marine invertebrate larvae. Chia and Rice, eds. Elsevier/North-Holland Biomedical Press. p. 49-63.

Zhang, X.-G., and Pratt B.R. 1994. New and extraordinary Early Cambrian sponge spicule assemblage from China. Geol. 22: 43-46.

Zimmer, R.L., and Woollacott R.M. 1977. Metamorphosis, ancestrulae, and coloniality in bryozoan life cycles. In: Biology of Bryozoans. R. M. Woollacott and R. L. Zimmer, eds. Academic Press, New York. p. 91-142.

Zimmer, R.L., and Woollacott R.M. 1983. Mycoplasma-like organisms: Occurrence with the larvae and adults of a marine bryozoan. Science. 220: 208-210.

6

Microbial Gene Transfer: An Ecological Perspective

John H. Paul

Department of Marine Science, University of South Florida,
140 Seventh Ave S., St. Petersburg, FL 33701, USA

Abstract

Microbial gene transfer or microbial sex is a means of exchanging loci amongst prokaryotes and certain eukaryotes. Historically viewed as a laboratory artifact, recent evidence from natural populations as well as genome research has indicated that this process may be a major driving force in microbial evolution. Studies with natural populations have taken two approaches-either adding a defined donor with a traceable gene to an indigenous community, and detecting the target gene in the indigenous bacteria, or by adding a model recipient to capture genes being transferred from the ambient microbial flora. However, both approaches usually require some cultivation of the recipient, which may result in a dramatic underestimation of the ambient transfer frequency. Novel methods are just evolving to study *in situ* gene transfer processes, including the use of green fluorescent protein (GFP)-marked plasmids, which enable detection of transferrants by epifluorescence microscopy. A transduction-like mechanism of transfer from viral-like particles produced by marine bacteria and thermal spring bacteria to *Escherichia coli* has been documented recently, indicating that broad host range transduction may be occurring in aquatic environments. The sequencing of complete microbial genomes has shown that they are a mosaic of ancestral chromosomal genes interspersed with recently transferred operons that encode peripheral functions. Archaeal genomes indicate that the genes for replication, transcription, and translation are all eukaryotic in complexity, while the genes for intermediary metabolism are purely bacterial. And in eukaryotes, many ancestral eukaryotic genes have been replaced by bacterial genes believed derived from food sources. Collectively these results indicate that microbial sex can result in the dispersal of loci in contemporary microbial populations as well as having shaped the phylogenies of microbes from multiple, very early gene transfer events.

Introduction

Gene transfer can generally be regarded as prokaryotic sex, or a mechanism to "mix genes" between prokaryotic community members. However, the process of prokaryotic gene transfer is dramatically different from sex as it occurs in dioecious higher organisms. First, it is usually not a component of the cellular life cycle in prokaryotes. This might be contested in the case of transformation, as the genes for competence development undergo activation at discreet times in the growth cycle of batch cultures (Smith *et al.*, 1981). Secondly, the frequency of transfer is usually quite low. The frequency of transfer is expressed as the number of transferrants (*i.e.* cells which represent the successful acquisition of a gene) divided by the number of total recipients (or donors) present. Very high transfer frequencies (usually conjugation) are 10^{-1} to 10^{-3} transferrants per recipient, while transfer in the environment can be as low as 10^{-8} to 10^{-9} (Frischer *et al.*, 1994, Jiang and Paul, 1998). Even though the transfer frequencies observed may be low, this does not mean that gene transfer has had very little impact on microbial communities. Low frequencies of transfer over geologic time is believed to be a major driving force in microbial evolution (Pennisi, 1998).

Unlike sex in higher organisms, only one or a very limited number of closely linked loci are transferred at any one particular time. Entire genomes are never recombined, as for diploid organism. This leads to "micro-evolution of loci". And certain loci are known to transfer more frequently than others.

Lastly, the limit on the nature or genetic relatedness of the participants does not have the same restrictions as that of dioecious organism. The "species concept" or producing fertile offspring has no meaning in prokaryotic sex. Indeed, the diversity in partners participating in prokaryotic sex is not even limited by kingdoms.

Gene transfer was regarded as a laboratory artifact for many years. The potential use of recombinant microbes in the environment stimulated gene transfer research in the 1980's. This required development of methods to try to detect gene transfer, both in the lab and in the environment. As a result, there have been three general approaches to investigating the how, when, and why of environmental gene transfer. The first is to add an easily detected gene to the environment and investigate acquisition by the indigenous flora (Frischer *et al.*, 1994). This is perhaps most satisfying in that one can detect the exogenous gene by probing or PCR (Williams *et al.*, 1997). The only caveat is to be certain that signal detected is from the gene in a recipient and not in the donor that was added (termed counter selection against the donor). The second is to add a recipient and detect transfer of an indigenous marker to the recipient. This is clearly more difficult because one often doesn't know what to look for in terms of phenotype or genotype. The third approach is to detect acquired genes in the genomes of environmental isolates. This is usually done by sequence analysis of loci in question and finding divergent sequences. This is satisfying because it affirms the importance of gene transfer. However, the questions of when, how, and by what mechanism are then merely the subject of speculation. None the less, it is in this arena that gene transfer is experiencing a revival in interest. The sequencing of entire microbial genomes (Fleischman *et al.*, 1995; Frazer *et al.*, 1995; Bult *et al.*, 1996) has shown that microbes are mosaics of acquired genes. In fact, the rooting of the universal tree is highly problematic because of multiple transfer events between the kingdoms.

In this review I have focused on gene transfer to indigenous organisms in the environment, paying less attention to laboratory and microcosm studies. Because much work has been done on conjugation and less on transduction, I emphasize some of the more startling findings coming from the latter field. Finally, how gene transfer has impacted our understanding of molecular phylogeny will be examined.

Mechanisms of Transfer

The three recognized mechanisms of prokaryotic gene transfer are transformation, conjugation, and transduction. These mechanisms should be viewed as a framework for conceptualizing gene transfer in the environment. However, other novel mechanisms may exist, and combinations of mechanism may occur in the environment.

Transformation was the first mechanism of gene transfer to be recognized (Griffith, 1928) and its discovery paved the way for understanding that DNA was the genetic material in all cells. Transformation involves the uptake and expression of genes encoded in extracellular DNA. Unlike conjugation and transduction, transformation is a normal, physiological function of certain bacteria and is mediated by chromosomal genes (Smith *et al.*, 1981). A distinction is made between natural transformation and artificial transformation in regard to competence, or the capability to uptake DNA. Naturally competent bacteria express competence at some time in their life cycle, whereas artificially induced competence, as used in plasmid transformation in *E. coli*, is the result of chemical or physical perturbation of the cell membrane/wall to enable covalently close plasmid penetration (Mandel and Higa, 1970).

Conjugation is a plasmid or transposon encoded mechanism of transfer that requires cell contact, first described by Lederburg and Tatum (1946). Because of their location on plasmids or transposons, the genes transferred in conjugation generally encode accessory functions such as antibiotic, UV, and heavy metal resistance or expanded metabolic capabilities (*i.e.* Xenobiotic degradation; Willets and Wilkins, 1984). Conjugation has often been viewed as the most promiscuous of the three transfer mechanisms because of the lesser restriction on similarity of recipient to donor imposed by transformation and transduction.

Transduction, which literally means "carrying over', is the process of gene transfer whereby a phage mistakenly packages some host DNA in the capsid and transfers it to another bacterium upon subsequent infection (Zinder and Lederburg, 1952). This process can result in either the transfer of a random fragment of the host genome or plasmid (termed generalized transduction) or, when a temperate phage is employed, specific genes which flank the place of prophage integration (Ackermann and DuBow, 1987). Although perhaps the least understood mechanism of environmental gene transfer, recent findings indicate that this process may be occurring more frequently in the marine environment than previously thought.

Other methods of gene transfer include capsduction, or gene transfer by a small phage-like structure (Joset and Guepsin-Michel, 1993), protoplast fusion (Matsushima and Baltz, 1986) and transposition.

Transformation

The potential for gene transfer by transformation in the environment was first demonstrated by Graham and Istock (1978) using chromosomal loci of *Bacillus subtilis* in sterile soil. Since these early studies, many reports have described transformation in sand/seawater microcosms, although most often with sterile sand and *B. subtilis* as recipient (Aardema *et al.*, 1983; Lorenz and Wackernagel, 1987; Lorenz and Wackernagel, 1994). These studies have indicated that DNA binds to sand, becoming protected from DNase, but still capable of causing transformation.

Copious amounts of high MW dissolved DNA in the marine environment (DeFlaun *et al.*, 1987; Paul *et al.*, 1987) suggested that transformation could be a viable mechanism of gene transfer. Gene transfer to a High Frequency of Transformation *Vibrio* was demonstrated using plasmid multimers and seawater microcosms (Paul *et al.*, 1991) The presence of the ambient microbial community in water column simulations either had no effect on the rate of transfer or inhibited transfer. Nutrients stimulated the transfer rate. These results suggested that nutrients facilitated transfer, probably by stimulating recipient growth, and by providing a preferred carbon source (over the transforming DNA) from the indigenous community. At low nutrient conditions, we hypothesized that the DNA was used as a C/N/P source by the indigenous flora, and was not available to transform the recipients. In sediments, the presence of the indigenous flora inhibited gene transfer, whereby sterile sediments enabled transfer to occur (Paul *et al.*, 1991). In soil, *Acinetobacter calcoaceticus* could only be transformed in nonsterile microcosms when excess nutrients were present (Nielsen *et al.*, 1997). Thus, it sees that studies with sterile sediments or soil do not yield relevant information concerning transfer in the marine environment.

The facilitation of natural transformation by cell contact which was DNase sensitive was demonstrated by Stewart *et al.* (1983) in *Pseudomonas stutzeri*. The intergeneric transfer of a small, non- conjugative plasmid from *E. coli* to *Vibrio* JT-1 (later identified as a *Pseudomonas* species) occurred and was shown to be DNase-sensitive (Paul *et al.*, 1992). Heat -killed donor cells were as efficient as living cells, and interestingly transfer in liquid yielded a 10-fold higher frequency of transfer than filter matings.

Environmental conditions in microcosms may reveal gene transfer properties of bacterial strains not revealed by culture studies using rich media. For example, *E. coli* has been known for some time not to be naturally competent, only being transformable by chemical treatment, osmotic shock, and electroporation. Using river water, spring water and mineral water, *E. coli* was shown to be naturally competent for the uptake of pUC18 DNA (Baur *et al.*, 1996). Competence seemed to be internally regulated, being greatest in exponential phase and least in stationary phase. This work argues for the potential for *E.coli* (and perhaps other coliform bacteria) to be transformed in natural environments.

Transfer to the Indigenous Community

The capability to measure transformation of DNA in natural microbial communities has challenged investigators for years. Nearly all such attempts have used closed microcosms. However, the work of Day and Fry (Williams *et al.*, 1992; Williams *et al.*, 1996) is unique in that an indigenous mercury resistance plasmid (pQM17) was used in the river from which it was isolated in open filter matings. Stones with filters containing the donors or DNA were placed next to stones containing recipients. Transfer occurred with frequencies ranging from 10^{-6} to 10^{-4} per recipient. A significant effect of temperature was observed, with no transfer below 10 °C.

Natural transformation of indigenous marine bacteria has been demonstrated by Frischer *et al.* (1994) using plasmid multimers of pQSR50. Plasmid multimers were used to ensure internal homology and thus provide a site for self-homologous recombination. The ambient microbial communities in water column samples (20 L) or coral mucus (1 L) were first concentrated to approx. 50 ml and 1 ml used in filter transformation assays. For sediments and bacteria in sponge tissue, the bacteria were extracted mechanically and then used in filter transformation assays. Transformation was assessed as expression of the antibiotic resistance genes encoded by the plasmid in combination with confirmation by molecular probing. However, certain environments contained indigenous marine bacteria that possessed sequences that hybridized with the probe used and any such environments were not considered further. Positive transformation was found in 5 of 13 experiments, with transfer frequencies ranging from 3.6×10^{-6} to 1.13×10^{-9}. In all cases of plasmid transfer to the natural community, restriction profiles from transformants were altered when compared to the parent plasmid. Differences in the recovered transformant plasmids were accounted for by difference in methylation compared to the parent plasmid, as well as some genetic rearrangement (Williams *et al.*, 1997). Thus, transfer to the indigenous flora, at least when considering plasmid DNA, can result in rearrangement and alteration of the DNA, contributing to plasmid and recipient evolution.

Conjugation

As mentioned, conjugation is often viewed as the most promiscuous of the gene transfer mechanisms, showing the least restriction in relatedness of host/recipients. Although conjugation has been observed in soils, insect larvae (Vila-Boas *et al.*, 1998) seed surfaces (Lilley and Baily, 1997; Lilley *et al.*, 1996), in raw salmon on a cutting board (Kruse and Sorum, 1994), in porcine feces (Kruse and Sorum, 1994), crown gall tumors (Zhang *et al.*, 1993) and a variety of other environments, few studies have focused on the indigenous flora, particularly in the marine environment.

A problem in the study of gene transfer in aquatic or novel environments is that conjugative plasmids used in such studies are of terrestrial origin and may not be representative of or behave like those indigenous to the environment in question. Dahlberg *et al.* (1997) isolated 95 mercury resistance conjugative plasmids from a variety of marine environments by transfer to a model recipient. There was a tremendous diversity in plasmids and 12 general structural groups were differentiated. Plasmids from different environments showed little similarity as determined by RFLP's. The plasmids were further-tested for similarity to known plasmid incompatibility groups using inc/rep probes (Couturier et al., 1988). Hybridization studies showed no similarity between any of the twelve groups of naturally occurring marine plasmids and the well characterized plasmids often used in gene transfer studies. This study shows the need to develop more environmentally relevant plasmids and probes for the study of conjugation in situ.

Sandaa and Enger (1994) noted transfer of the highly promiscuous plasmid pRAS1 from the fish pathogen Aeromonas salmonicida to the indigenous flora in marine sediments. If oxytetracycline was added as a selective pressure, transfer frequencies increased to 3.4×10^{-1} transconjugants per recipient and 3.6 transconjugants per donor, perhaps the highest reported gene transfer frequency for any environmental gene transfer study. The plasmid was transferred to a variety of biotypes as determined by phenotypic characterization of the recipients. The transfer frequencies expressed per recipient are undoubtedly overestimates, because recipients were enumerated as cultivatable CFU's and not direct counts. None-the-less, this work illustrates the promiscuity of certain conjugative plasmids, and the ease at which antibiotic resistances can spread through natural marine bacterial communities.

Perhaps one of the most elegant demonstrations of conjugation to the indigenous marine flora was the work of Dahlberg et al. (1998). These investigators used the conjugative plasmid pBF1 which contains the green fluorescent protein (GFP) gene in Pseudomonas putida as a donor. This gene is not expressed in the donor because of a chromosomal repressor. Gene transfer was noted when cultivated recipients were used as well as the natural marine population. Transfer to the indigenous population could be observed directly by epifluorescence microscopy, without the need of cultivation of transconjugants. Transfer frequencies ranged from 2×10^{-6} to 1.4×10^{-4}, with transfer highest in filter mating compared to bulk matings. The plasmid used also encoded for mercury resistance, but selection in mercury did not facilitate or enhance transfer frequencies. Thus, selective pressure may or may not enhance transfer to the indigenous population.

Transduction

The discovery of the tremendous number of viruses in aquatic environments (Berg et al., 1989; Proctor and Fuhrman, 1990; Paul et al., 1991) has raised the issue of these as potential gene transfer vectors. The elegant studies of Miller and coworkers with microcosms in lakes using the P. aeruginosa system (Miller et al., 1977; Saye et al., 1987; Ripp and Miller, 1995) has demonstrated the potential for transduction of both plasmid and chromosomal markers. Frequency of transduction was approximately 10^{-7} per recipient.

Jiang and Paul (1998) described plasmid transduction of a lysogenic marine phage host system, ϕHSIC/HSIC. Concomitant with acquisition of plasmid DNA was the acquisition of a portion of the viral genome, as indicated by molecular probing with a 100 bp gene fragment of the transducing phage. Transfer frequencies were quite low (10^{-9} to 10^{-7}).

To investigate transfer to the indigenous marine population by transduction, Jiang and Paul (1998) produced lysates from host bacteria that contained the plasmid pQSR50. The lysates were then added to concentrated indigenous marine microbial communities. One phage, ϕD1B, yielded transductants of the natural population. Transductants were verified by probing with the Tn5 sequences. As found for natural transformation of plasmid DNA, restriction profiles were modified in transductants. Amplification of the transductants using plasmid specific primers indicated correct amplification of the kanamycin resistance gene but that a restriction site was missing from another portion of the plasmid. This indicates that , as with natural plasmid transformation, some rearrangement/recombination of the plasmid DNA occurred by the natural community.

Perhaps the most compelling and unique evidence for transduction comes from the work of Chiura and coworkers (1997, 1998). These investigators observed VLP (virus-like particle) production in five marine bacterial cultures, with the ratio of viruses to host approximately 1.0. These VLP's could transfer amino acid prototrophy to an auxotrophic *E. coli* strain. The viruses were lethal to the *E. coli*. The gene transfer frequency was between 2.6×10^{-3} to 3.58×10^{-5} per VLP. These results suggested that some type of generalized transduction was occurring, and that such gene transfer could occur using a very broad host range of donors. Unfortunately, because there were no probes available for the loci transferred, it was not possible to verify gene acquisition by probing.

These results were quite surprising because the viral-host interaction is viewed to be quite specific. However, recent findings on the host range of marine Myoviridae indicates that some phages can be quite broad host range (Wichels *et al.*, 1998). Therefore, broad host range transduction may be an important mechanism of horizontal transfer in the marine environment. Supporting this concept were other findings of Chiura *et al.* (1998) using VLP's that were produced by natural communities of geothermal hot spring bacteria. These VLP's could transduce *E. coli* to prototrophy, with an average transfer frequency of 10^{-6}/VLP. VLP's were purified from natural populations without cultivation and these also caused transfer to prototrophy.

The mechanism of transfer to prototrophy was not known but believed to involve simple complementation of the mutant. Collectively these results indicate that gene transfer by transduction can occur across wide taxonomic boundaries in the marine and hot spring environments.

Gene Transfer and Microbial Evolution

Not too long ago, attributing an unusual genomic finding to a lateral transfer event would have been met with general skepticism. Irregularities in phylogenetic trees often were thought to be caused by errors in tree construction. However, complete genome analysis has made the concept of gene transfer actually indispensable to understanding microbial evolution and phylogeny in general.

"Lateral gene transfer is not just a molecular phylogenetic nuisance supported in evidence by a few anecdotal cases. Instead, it is a major force, at least in prokaryotic evolution." Doolittle and Logsdon, 1998.

The molecular evidence for gene transfer is overwhelming and leading to a general acceptance of this concept. For example, Lawrence and Roth (1996) concluded from the examination of the bacterial genome that it was a mosaic of ancestral chromosomal genes interspersed with some recently transferred operons that encode peripheral functions. The successful transfer of complete functions is facilitated by genes which are in clusters, opposed to those that are spread out throughout the genome. Thus, functions encoded by genes within close proximity of one another have a greater likelihood of being transferred than those which are on different parts of the chromosome.

The sequencing of complete genomes of archaea, bacteria, and eukaryotes has clearly "shaken the tree of life" (Pennisi, 1998). The concept that organelles in eukaryotes arose from endosymbiotic capture (Margulis, 1970) actually represents a large scale gene transfer event. rRNA phylogeny had endorsed a tree that included a primordial progenitor that gave rise to bacteria, and then later a second branch which diverged into eucarya and archaea. However, new evidence suggests that both bacterial and archeal genes are showing up in eukaryotes (Pennisi, 1998). Additionally, archaea have many eukaryotic traits (Doolittle and Logsdon, 1998). It now appears that early organisms stole genes from branch to branch, either from food or swapping genes with neighbors. Thus, it is difficult to make sense of phylogenies because of all the gene swapping, and different phylogenies are obtained depending on what genes are examined. For example, the archaea look fairly homogeneous in terms of 16S rRNA phylogenies, but the CTP synthetase gene spread archaea amongst all other organisms evaluated (Penninsi, 1998). The spirochaete *Treponema palladium* contains two Archeal ATPases.

Archaea are now recognized as having mixed heritage. The genes for replication, transcription, and translation are all eukaryote-like in complexity (Doolittle and Logsdon, 1998). However, the genes of intermediary metabolism are purely bacterial. The Archaeal genome sequencing papers indicate that the genes for replication, transcription, and transduction form an enduring cellular hardware, whereas the other genes for biochemical functions are a type of software, and prone to change.

A major mechanism for acquisition of genes by eukaryotes may have been grazing on prokaryotes and/or eukaryotes ("you are what you eat"; Doolittle, 1998). That is, over time, there has been a tendency to replace true eukaryotic or archeal genes in eukaryotes with those coming in from the food source. Although such transfer may be infrequent, there is a constant influx of genes, and once a eukaryotic gene is lost, it is lost for good. The replacement of archaeal genes in eukaryotes was thought to occur quite early in the evolution of eukaryotes, and that archaeal genes are found at all is surprising (Doolittle, 1998).

Perhaps one of the most unique horizontal transfer events to have occurred was from bacteria to vertebrates. The rearrangement of antibody and T-cell receptors is indispensable to the vertebrate immune response, and is conferred by the recombination activating genes I and II (RAG I and II) (Bernstein *et al.*, 1996) An examination of one of the most primitive jawed vertebrates, the carnachine sharks, showed that the sequence of RAG I and RAG II were most closely related to the integrase family of genes and the integration factors, respectively, of bacterial site-specific recombination systems. The apparent "Big Bang" of ancestral marine systems that occurred in jawed vertebrates was apparently due to transfer of microbial site specific recombinases.

Concluding Remarks

Our understanding of microbial gene transfer has evolved through the use of the three approaches described in the introduction. If understanding the mechanism of gene transfer is important, then using a defined donor system and the indigenous population as recipient is the best approach. Gene transfer can be confirmed by detection of the transferred DNA by probing. If the capability for transfer to occur is important and the presence of donors needs to be known, then adding a universal recipient to the system is the best approach. If one doesn't care at all about the mechanism but that only a transfer has occurred throughout the evolutionary history of the organism, then genome analysis is the appropriate approach.

The mechanisms of gene transfer, described in detail for type strains, have provided a framework for the understanding of environmental gene transfer. Sterile microcosm studies with known donors and recipients have provided some valuable information on the potential for transfer to occur, but may be misleading because of the complications that the indigenous flora impose. Additionally, terrestrially-derived models of gene transfer may have no validity in marine or other environments. There is a need for the isolation and description of marine gene transfer systems, which may reveal novel mechanisms of transfer. Indeed, the division of the three mechanisms of gene transfer (transformation, conjugation, and transduction) may diffuse into grey areas of overlap when describing transfer in natural environments. It can be envisioned that a phage may lyse a bacterium containing a conjugative plasmid, which may transform a bacterium, and then be transferred by conjugation to second recipient. Transducing phages could conceivably carry conjugative plasmids or transposons.

In view of all the evidence for gene transfer, perhaps the focus now should not be on "does gene transfer occur?" or "what genes are the result of lateral transfer" but rather "what prevents certain loci from being transferred?" How does an evolutionary anchor get established, or what determines what becomes genetic hardware vs. genetic software? These are all questions worthy of study in the understanding of microbial sex in the new millenium.

References

Aardema, B.W., Lorenz, M.G., and Krumbein, W.E. 1983. Protection of sediment adsorbed transforming DNA against enzymatic inactivation. Appl. Environ. Microbiol. 46: 417-420.

Ackermann, H.-W., and DuBow, M.S. 1987. Viruses of Prokaryotes. Vol. I. General properties of bacteriophages. CRC Press, Inc. Boca Raton, FL.

Baur, B., Hanselmann, K., and Schlimme, W. 1996. Genetic transformation in freshwater: *Escherichia coli* is able to develop natural competence. Appl. Environ. Microbiol. 62: 3673-3678.

Bergh, O., Borsheim, K.Y., Bratbak, G., and Heldal, M. 1989. High abundances of viruses found in aquatic environments. Nature. 340: 467-468.

Bernstein, R.M., Schulter, S.F., Bernstein, H., and Marchalonis, J.J. 1996. Primordial emergence of the recombination activating gene 1 (RAG1): Sequence of the complete shark gene indicates homology to microbial integrases. Proc. Nat. Acad. Sci. USA. 93: 9454-9459.

Bult, C.J., White, O., Olsen, G.J., Zhou, L., Fleischmann, R.D., Sutton, G.G., Blake, J.A., FitzGerald, L.M., Clayton, R.A., Gocayne, J.D., Kerlavage, A.R., Cougherty, B.A., Tomb, J.-F., Adams, M.D., Reich, C.I., Overbeek, R., Kirkness, E.F., Weinstock, K.G., Merrick, J.M., Glodek, A., Scott, J.L., Geoghagen, S.M., Weidman, J.F., Fuhrmann, J.L., Nguyen, D., Utterback, T.R., Kelley, J.M., Peterson, J.D., Sadow, P.W., Hanna, M.C., Cotton, M.D., Roberts, K.M., Hurst, M.A., Kaine, B.P., Borodovsky, M. Klenk, H-.P., Fraser, C.M., Smith, H.O., Woese, C.R., and Venter, J.C. 1996. Complete genome sequence of the methanogenic archaeon, *Methanococcus jannaschii*. Science. 273: 1058-1073.

Chiura, H.X. 1997. Generalized gene transfer by virus-like particles from marine bacteria. Aquatic Microbial Ecology. 13: 75-83.

Chiura, H.X., Kato, K., Hiraishi, A., Maki, Y. 1998. Gene transfer mediated by virus of novel thermophilic bacteria in hot spring sulfur-turf microbial mats. Eighth International Symposium on Microbial Ecology (ISME-8). Program and Abstracts, p. 124.

Couturier, M.F., Bex, P., Bergquist, L., and Maas, W.K. 1988. Identification and classification of bacterial plasmids. Microbiol. Revs. 52: 375-395.

Dahlberg, C., Bergstrom, M., and Hermansson, M. 1998. *In situ* detection of high levels of horizontal plasmid transfer in marine bacterial communities. Appl. Environ. Microbiol. 64: 2670-2675.

Dahlberg, C., Linberg, C., Torsvik, V.L., and Hermansson, M. 1997. Conjugative plasmids isolated from bacteria in marine environments show vaious degrees of homology to each other and are not closely related to well-characterized plasmids. Appl. Environ. Microbiol. 63: 4692-4697.

DeFlaun, M.F., Paul, J.H., and Jeffrey, W.H. 1987. Distribution and molecular weight of dissolved DNA in subtropical estuarine and oceanic environments. Mar. Ecol. Prog. Ser. 38: 65-73.

Doolittle, W.F. 1998. You are what you eat: a gene transfer ratchet could account for bacterial genes in eukaryotic nuclear genomes. Trends Genet. 14: 307-311.

Doolittle, W.F., and Logsdon, J.M., Jr. 1998. Archaeal genomics: Do archaea have a mixed heritage? Curr. Biol. 8: R209-R211.

Fleischman, R.D., Adams, M.D., White, O., Clayton, R.A., Kirkness, E.F., Kerlavage, A.R., Bult, C.J., Tomb, J.-F., Dougherty, B.A., Merrick, J.M., McKenney, K., Sutton, G., FitzHugh, W., Fields, C., Gocayne, J.D., Scott, J., Shirley, R., Liu, L.-I., Glodek, A., Kelley, J.M., Weidman, J.F., Phillips,

C.A., Spriggs, T., Hedblom, E., Cotton, M.D., Utterback, T.R., Hanna, M.C., D.T. Nguyen, D.T., Saudek, D.M., Brandon, R.C., Fine, L.D., Fritchman, J.L., Fuhrmann, J.L., Geoghagen, N.S.M., Gnehm, C.L., McDonald, L.A., Small, K.V., Fraser, C.M., Shith, H.O., and Venter, J.C. 1995. Whole-genome random sequencing and assembly of *Haemophilis influenzae* Rd. Science. 269: 496-512.

Frazer, C.M., Gocayne, J.D., White, O., Adams, M.D., Clayton, R.A., Fleishmann, R.D., Bult, C.J., Kerlavage, A.R., Sutton, G., Kelley, J.M., Fritchman, J.L., Weidman, J.F., Small, K.V., Sandusky, M., Fuhrmann, J., Nguyen, D., Utterback, T.R., Saudek, D.M., Phillips, C.A., Merrick, J.M., Tomb, J.-F., Dougherty, B.A., Bott, K.F., Hu, P.-C., Lucier, T.S., Peterson, S.N., Smith, H.O., Hutchison III, C.A., and Venter, J.C. 1995. The minimal gene complement of *Mycoplasma genitalium*. Science. 270: 397-410.

Frischer, M.E., Stewart, G.J., and Paul, J.H. 1994. Plasmid transfer to indigenous marine bacterial populations by natural transformation. FEMS Microbiol. Ecol. 15: 127-136.

Graham, J.B., and Istock, C.A. 1978. Genetic exchange in *Bacillus subtilis* in soil. Mol. Gen. Genet. 166: 287-290.

Griffith, F. 1928. Significance of pneumonococcal types. J. Hyg. 27: 113-159.

Jiang, S.C., and Paul, J.H. 1998. Gene transfer by transduction in the marine environment. Appl. Environ. Microbiol. 64: 2780-2787.

Joset, F., and Guespin-Michel, J. 1993. Prokaryotic Genetics: Genome Organization, Transfer, and Plasticity. Blackwell Scientific, Oxford.

Kruse, H., and H. Sorum. 1994. Transfer of multiple drug resistance plasmids between bacteria of diverse origins in natural microenvironments. Appl. Environ. Microbiol. 60: 4015-4021.

Lawrence, J.G., and Roth, J.R. 1996. Selfish operons: Horizontal transfer may drive the evolution of gene clusters. Genet. 143: 1843-1860.

Lederberg, J., and Tatum, E.L. 1946. Novel genotypes in mixed cultures of biochemical mutants of bacteria. Cold Spring Harbor Symp. Quant. Biol. 16: 413-443.

Lilley, A.K., and Bailey, M.J. 1997. The acquisition of indigenous plasmids by a genetically marked pseudomonad population colonizing the sugar beet phytosphere is related to local environmental conditions. Appl. Environ. Microbiol. 63: 1577-1583.

Lilley, A.K., Bailey, M.J., Day, M.J., Fry, J.C. 1996. Diversity of mercury resistance plasmids obtained by exogenous isolation from the bacteria of sugar beet in three successive years. FEMS Microbiol. Ecol. 20: 211-227.

Lorenz, M.G., and Wackernagel, W. 1987. Adsorption of DNA to sand and variable degradation rates of adsorbed DNA. Appl. Environ. Microbiol. 53: 2948-2952.

Lorenz, M.G., and Wackernagel, W. 1994. Bacterial gene transfer by natural genetic transformation in the environment. Microbiol. Rev. 58: 563-602.

Margulis, L. 1970. Origin of eukaryotic cells. Yale University Press, New Haven, CT.

Mandel, M., and Higa, A. 1970. Calcium-dependent bacteriophage DNA infection. J. Mol. Biol. 53: 159-162.

Matsushima, P., and Baltz, R.H. 1986. Protoplast fusion. In: Manual of

Industrial Microbiology and Biotechnology. A.L. Demain and N.A. Solomon, eds. American Society for Microbiology, Washington, DC. p. 170-183.

Miller, R.V., Pemberton, J.M., and Clark, A.J. 1977. Prophage F116: Evidence for extrachromosomal location in *Pseuomonas aeruginosa* strain PAO. J. Virol. 22: 844-847.

Nielsen, K.M., van Weerelt, Margareta D.M., and Berg, T.N. 1997. Natural tranaformation and availability of transforming DNA to *Acinetobacter calcoaceticus* in soil microcosms. Appl. Environ. Microbiol. 63: 1945-1952.

Paul, J.H., Thurmond, J.M., Frischer, M.E., and Cannon, J.P. 1992. Intergeneric natural plasmid transformation between *E. coli* and a marine *Vibrio* species. Mol. Ecol. 1: 37-46.

Paul, J.H. , Frischer, M.E., and Thurmond, J.M. 1991. Gene transfer in marine water column and sediment microcosms by natural plasmid transformation. Appl. Environ. Microbiol. 57: 1509-1515.

Paul, J.H., Jeffrey, W.H., and DeFlaun, M.F. 1987. The dynamics of extracellular DNA in the marine environment. Appl. Environ. Microbiol. 53: 170-179.

Paul, J.H., Jiang, S.C., and Rose, J.B. 1991. Concentration of viruses and dissolved DNA from aquatic environments by vortex flow filtration. Appl. Environ. Microbiol. 57: 2197-2204.

Pennisi, E. 1998. Genome data shake tree of life. Science. 280: 672-674.

Proctor, L.M and Fuhrman, J.A. 1990. Viral mortality of marine bacteria and cyanobacteria. Nature. 343: 60-62.

Ripp, S., and Miller, R.V. 1995. Effects of suspended particulates on the frequency of transduction among *Pseudomonas aeruginosa* in a freshwater environment. Appl. Environ. Microbiol. 61: 1214-1219.

Sandaa, R.-A., and Enger, O. 1994. Transfer in marine sediments of the naturally occurring plasmid pRAS1 encoding multiple antibiotic resistance. Appl. Environ. Microbiol. 60: 4234-4238.

Saye, D.J., Ogunseitan, O., Sayler, G.S., and Miller, R.V. 1987. Potential for transduction of plasmids in a natural freshwater environment. Effect of plasmid donor concentration and a natural microbial community on transduction in *Pseudomonas aeruginosa*. Appl. Environ. Microbiol. 53: 987-995.

Smith, H.O., D.B. Danner, and R.A. Deich. 1981. Genetic transformation. Annu. Rev. Biochem. 50: 41-68.

Stewart, G.J., Carson, C.A., and Ingraham, J.L. 1983. Evidence for an active role of donor cells in natural transformation in *Pseudomonas stutzeri*. J. Bacteriol. 156: 30-35.

Vilas-Boas, C.F.L.T., Vilas-Boas, L.A., Lereclus, D., and Arantes, O.M.N. 1998. *Bacillus thuringiensis* conjugation under environmental conditions. FEMS Microbiol. Ecol. 25: 369-374.

Wichels, A., Biel, S.S., Gelderblom, H.R., Brinkhoff, T., Muyzer, G., and Schutt, C. 1998. Bacteriophage diversity in the North Sea. Appl. Environ. Microbiol. 64: 4128-4133.

Willets, N., and Wilkins, B. 1984. Processing of plasmid DNA during bacterial conjugation. Microbiol. Rev. 48: 24-41.

Williams, H.G., Benstead, J., Frischer, M.E., and Paul, J.H. 1997. Alteration in plasmid DNA following natural transformation to populations of marine bacteria. Mol. Mar. Bio. Biotechnol. 6: 238-247.

Williams, H.G., Day, M.J., and Fry, J.C. 1992. Natural transformation on agar and in river epilithon. In: Gene Transfers and Environment. M.H. Gauthier, ed. Springer-Verlag, Berlin. p. 69-76.

Williams, H.G., Day, M.J., Fry, J.C., and Stewart, G.J. 1996. Natural transformation in river epilithon. Appl. Environ. Microbiol. 62: 2994-2998.

Zhang, L., Murphy, P.J., Kerr, A., and Tate, M.E. 1993. Agrobacterium conjugation and gene regulation by N-acyl-L-homoserine lactones. Nature. 362: 446-448.

Zinder, N.D., and Lederberg, J. 1952. Genetic exchange in *Salmonella*. J. Bacteriol. 64: 679-699.

7

Surface Induced Gene Expression in *Vibrio parahaemolyticus*

Linda McCarter

Microbiology Department, University of Iowa,
Iowa City, Iowa 52242, USA

Abstract

Vibrio parahaemolyticus is a ubiquitous marine bacterium and human pathogen. The organism possesses multiple cell types appropriate for life under different circumstances. The swimmer cell, with a single polar flagellum, is adapted to life in liquid environments. The polar flagellum is powered by the sodium motive force and can propel the bacterium at fast speeds. The swarmer cell, propelled by many proton-powered lateral flagella, can move through highly viscous environments, colonize surfaces, and form multicellular communities which sometimes display highly periodic architecture. Signals that induce differentiation to the surface-adapted cell type are both physical and chemical in nature. Surface-induced gene expression may aid survival, whether attached to inanimate surfaces or in a host organism. Genetic rearrangements create additional phenotypic versatility for the organism and is manifested as variable opaque and translucent colony morphotypes. Discovery that a LuxR homolog controls the opaque cell type implicates intercellular signaling as an additional survival strategy. The alternating identities of *V. parahaemolyticus* may play important roles in attachment and detachment, how bacterial populations adapt to growth on surfaces, form structured communities, and develop biofilms.

Introduction

Occupying a variety of niches, *Vibrio parahaemolyticus* is a common bacterium in marine and estuarine environments. It can exist planktonically or attached to submerged, inert and animate surfaces, including suspended particulate matter, zooplankton, fish and shellfish (Kaneko and Colwell, 1973,

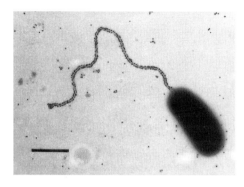

Figure 1. Electron Micrograph of a Swimmer Cell
The swimmer cell was reacted with rabbit
antiserum prepared against whole cells and
15nm protein A-bound colloidal gold particles.
Bar indicates approximately 1 μm.

1975). This organism is recognized as a major, worldwide cause of gastroenteritis, particularly in areas of the world where seafood consumption is high such as Southeast Asia (Joseph *et al.,* 1982). It is an emerging pathogen in North America. In 1997, a large outbreak of *V. parahaemolyticus* food poisoning, attributed to raw oyster consumption, occurred along the Pacific coast (CDC, 1998). Thus, *V. parahaemolyticus* seems suited to multiple lifestyles: a planktonic, free-swimming state and a sessile existence within a microbial community attached, for example, to shellfish in a commensal relationship, to the bottoms of boats or other surfaces in the ocean (causing biofouling), or in a host organism (causing pathogenesis). What are the survival strategies that allow this bacterium to adapt to life in dilute liquid environments and to life on surfaces or in biofilms?

The Swimmer Cell

The free-living form of *V. parahaemolyticus*, the swimmer cell (Figure 1), is well-suited for locomotion in liquid environments. The rod-shaped bacterium is efficiently propelled by a single polar flagellum. Energy to power the flagellum is derived from the sodium motive force (Atsumi *et al.,* 1992). Possession of sodium energetics has an advantage in the marine environment because the pH of seawater is approximately 8.0 (Kogure,

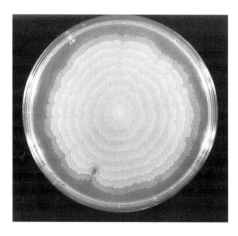

Figure 2. Swarming Colony on Rich Medium after Overnight Incubation at 30 °C.

1998). Sodium-driven flagellar motors are remarkably fast. In liquid medium with 300mM NaCl, the swimming speed of the bacterium is approximately 60μm per sec. Sodium-powered rotation of the polar flagellum of the closely related bacterium *Vibrio alginolyticus*, which swims at an equivalent speed in liquid, has been clocked using laser-darkfield microscopy at rates as fast as 1,700 r.p.s. (Magariyama *et al.*, 1994).

The polar organelle is a complex flagellum. There are six polar flagellin genes, organized in two loci (McCarter, 1995; GenBank Accession U12816 and U12817). Moreover, the polar flagellum is sheathed by what appears to be an extension of the cell outer membrane (Allen and Baumann, 1971). The mechanism of how a sheathed flagellum rotates has not been elucidated. Potentially, the flagellar filament could rotate within the sheath, or the sheath and filament could rotate as a unit (Fuerst, 1980). The flagellum plays a key role in initial adsorption of bacteria to surfaces. A plethora of studies with a variety of bacteria have shown that motility is important for adhesion as well as pathogenesis (O'Toole and Kolter, 1998; Otteman and Miller, 1997; Pratt and Kolter, 1998). As a propulsive organelle, it brings the bacteria into close proximity with surfaces and perhaps aids in overcoming negative electrostatic interactions. In fact, studies have shown that the faster the swimming, the greater the adhesion to glass (Kogure *et al.*, 1998). Flagellar sheaths allow specific interaction between a bacterium and a surface (Sjoblad and Doetsch, 1982; Sjoblad *et al.*, 1983). Certainly, it seems clear that the sheath extends the surface area of the bacterium. Cell-surface components, including potential adhesins, may be differentially distributed or available on the flagellum and cell body. The flagellum in Figure 1 is studded with colloidal gold particles, after immunogold-labeling with rabbit antiserum prepared against whole cells. In contrast, there is little gold-labeling of the cell body, suggesting that antigens are presented differently by the cell body and the sheathed flagellum.

The Swarmer Cell

Growth on surfaces or in viscous environments induces differentiation to the swarmer cell. Septation ceases and the cell elongates, usually up to 30 μm in length. Induction of a second motility system, the lateral system, leads to elaboration of numerous peritrichously arranged flagella. The swarmer cell is adapted for movement on surfaces or through highly viscous environments. The polar flagellum performs poorly in medium of high viscosity while lateral flagella efficiently propel the bacterium in highly viscous environments. The speed of laterally flagellated cells is unaffected, remaining constant at about 25μm per second, on addition of the long branched-chain polymer polyvinylpyrrolidone (10% PVP-360) which increases viscosity to 10 centipoise (cP). In comparison, polarly propelled swimming motility decreases from 60 to less than 15 μm per second under this condition (Atsumi *et al.*, 1992). It has been shown for *V. alginolyticus* that lateral flagella can work at viscosities as high as 100 cP (Atsumi *et al.*, 1996).

Lateral flagella are not sheathed, and the filament is polymerized from a single flagellin subunit (McCarter and Wright, 1993). The flagella are very sensitive to mechanical shearing (Allen and Baumann, 1971). So many flagella are produced per cell and sloughed, or broken off, that detached flagellar aggregates are readily visualized in the light microscope as giant, coiled bundles (Belas *et al.,* 1986; Ulitzur, 1974). These flagella are entirely distinct from the polar flagellum. Genetic experiments suggest that there are no shared structural components among the two motility systems. Mutants unable to swarm retain swimming motility and vice versa (McCarter *et al.,* 1988). Flagellar motility systems are typically encoded by at least 40 genes; therefore, two very large, distinct gene sets encoding the motility systems exist. When grown on a surface, the organism assembles two distinct types of flagella. What this means in terms of signals for flagellar export and assembly will be interesting to elucidate. Not only do the lateral and polar flagella differ at the structural level, but also are powered by different energy sources. Although in each case, energy is derived from the electrochemical transmembrane potential, the coupling ions differ. Lateral flagella are powered by the flow of protons through the motor in contrast to the sodium-driven polar motors (Atsumi *et al.,* 1992).

When inoculated on solidified medium, differentiation to the swarmer cell allows the bacterium to swarm, or move, over surfaces or through viscous environments, leading to colonization of surfaces (Figure 2). The movement is very vigorous. The percent agar over which swarming will occur is high (up to 2.0%), and the rate of radial expansion is rapid, for example, on 1.5% agar at 30°C swarming will progress at a rate of 5mm/hour. A number of bacteria are known to swarm. Although some show mixed flagellation like *V. parahaemolyticus* with distinct polar and peritrichous organelles (*e.g.,* *V. alginolyticus, Rhodospirillum centenum,* and *Azospirillum brasilense*; Allen and Baumann, 1971; Jiang *et al.,* 1998; Moens *et al.,* 1996), others possess single, peritrichous flagellar systems. In general the extent of hyperflagellation of the swarm cell correlates with successfulness of swarming. *V. parahaemolyticus* and *Proteus mirabilis* are superior swarmers and can have hundreds of flagella per cell. In comparison, *Serratia* species, *Escherichia coli,* and *Salmonella typhimurium* swarm under a more restricted set of conditions with agar concentrations at 0.5 – 0.8 % (Harshey, 1994; Harshey and Matsuyama, 1994). Movement for some of these swarming organisms has been shown to also require production of extracellular molecules capable of altering surface tension. Surfactant-like lipopeptides are produced by *S. marcesans* and *S. liquefaciens* (Matsuyama *et al.,* 1992; Lindum *et al.,* 1998). Rapid spreading of *P. mirabilis* requires production of extracellular capsular polysaccharide (Gygi *et al.,* 1995). Extracellular agents acting as swarming facilitators have not been demonstrated for *V. parahaemolyticus.* It may be that the copious production of fragile flagella substitutes to reduce frictional drag.

The concentric ring patterns or terraces, which are characteristic of the periodic swarming of *Proteus* species (Rauprich *et al.,* 1996), develop as a result of alternating cycles of active swarming and consolidation. During the consolidation period, swarmer cells divide to produce short cells, which eventually differentiate to the swarmer cells that initiate a new wave of swarming. Although the phenomenon has been studied extensively in *P. mirabilis*, this type of behavior is not restricted to *Proteus* species. Under certain conditions (specifically, swarm plates non-optimally conducive to

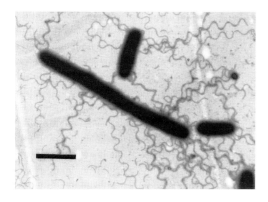

Figure 3. Electron Micrograph of a Constitutive Swarmer Cell
Grown in liquid and negatively-stained with 0.5 % phosphotungstic acid. Bar indicates approximately 3 μm.

swarming such as those made with Luria broth rather than Heart Infusion broth), certain *V. parahaemolyticus* strains produce the highly developed, periodic architecture shown in Figure 2. Detection of pattern formation may require a special balance between the rates of reproduction and movement of the cells as well as the particular surface properties of the substratum (i.e., surface tension).

Physical Signaling: The Polar Flagellum as Tactile Sensor

In addition to its role as a propulsive organelle and its potential for aiding attachment, the polar flagellum acts as a sensor. Polar flagellar function is coupled to expression of the swarmer cell gene system. The polar flagellum is produced constitutively, irrespective of liquid or surface-associated growth. All conditions that slow down polar flagellar rotation, lead to swarmer cell induction. Such conditions include increasing viscosity or using antibodies to inhibit flagellar function (Belas *et al.,* 1986, McCarter *et al.,* 1988). Physical conditions that impede flagellar rotation seem to act as a signal. Polar flagellar function can be perturbed in other ways. Genetic interference with polar function affects swarmer cell gene expression (McCarter *et al.,* 1988). All swimming-defective, transposon-generated mutants constitutively express swarmer cell genes when grown in liquid (Figure 3). The motor itself can be slowed down by using the sodium-channel-blocking drug phenamil (Kawagishi *et al.,* 1996). Just as increasing viscosity leads to swarmer cell differentiation, increasing concentrations of phenamil lead to decreasing swimming speed and concomitant induction of swarmer cell gene expression. Thus, the polar flagellum seems to act as a mechanosensor: interference with flagellar rotation signals swarmer cell differentiation. How signaling is transduced to program gene expression is not known. Current work involves dissecting the architecture of the sodium-driven motor of *V. parahaemolyticus* (McCarter, 1994a, 1994b; Jaques *et al.,* 1999).

Cell Division and the Long Cell Phenotype

One of the initial events after transfer to a surface is inhibition of cell division, and as a consequence, swarmer cells are characteristically very long.

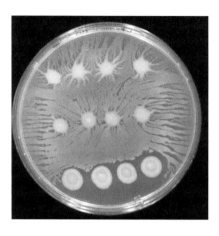

Figure 4. Swarming Colonies on Minimal Medium Galactose was the carbon source, 5-day incubation at 30°C. The strain inoculated in the bottom row carries a defect in the lateral motility system; therefore, it is unable to swarm.

Differentiation is transient. Since, swarmer cells must grow, i.e., divide, for the swarm colony to expand, the cell must have a mechanism to escape from the block to cell division. Thus, the inhibition of cell division during the swarm cell cycle must be carefully regulated for prolonged repression of cell division would be a terminal event.

Strains in one class of swarming constitutive mutants possess defects in *lonS*, a gene that codes for a homolog of the *E. coli* Lon protease (Stewart *et al.*, 1997). *E. coli lon* mutants were originally isolated as a class of UV-sensitive mutants (Gottesman, 1996). The role of Lon in *E. coli* is multifunctional, e.g., it targets the degradation of a transcriptional regulatory protein controlling production of extracellular polysaccharide (*cps*) as well as the cell division inhibitor SulA. SulA is induced by UV exposure as part of the SOS DNA repair response. Wild-type cells form long filaments after UV exposure, but with time SulA is degraded by Lon and the filaments resolve. Mutants with defects in *lon* cannot recover from elongation.

The *V. parahaemolyticus* counterpart closely resembles *E. coli lon* and substitutes functionally, by complementing *E. coli* mutants to restore UV resistance and *cps* regulation. In addition, *lonS* mutants in *V. parahaemolyticus* are more UV sensitive than the wild-type strain. It is attractive to hypothesize the existence of swarmer-cell specific cell division inhibitor. LonS could act as a policeman to keep in check the important cell division and regulatory proteins that mediate surface sensing. Potential regulatory targets for LonS could include transcriptional activators controlling the lateral flagellar gene system or extracellular polysaccharide.

Chemical Signaling

Differentiation to the swarmer cell requires considerable commitment in terms of cellular economy, including energy expenditure and gene expression. As a result, the developmental switch seems tightly controlled and multiple environmental stimuli are essential for cueing development. One cue that has been determined is iron starvation. Starvation signals seem an appropriate cue because the availability of nutrients or the diffusion of nutrients may be limiting for cells in a community attached to a substratum.

Development to the swarmer requires both iron limitation and perturbation of flagellar function (McCarter and Silverman, 1989). Nutrient deprivation and surface signaling may ensure detection of the specific conditions for which swarming is appropriate, e.g., a dense community of sessile cells.

Although others kinds of signals may also be important, *V. parahaemolyticus* can be induced to swarm on minimal medium (Figure 4; McCarter, 1998). This has not been shown to be the case for other bacteria where amino acids are implicated as swarming signals. Glutamine has been reported as essential to induce swarming of *P. mirabilis* on minimal medium (Allison *et al.,* 1993). Casamino acid supplementation is necessary for *S. liquefaciens* (Eberl *et al.,* 1996b); however, failure to swarm on minimal medium could also represent a growth rate barrier and not the lack of a specific inducing signal.

In Figure 4, the swarming colonies in the top and middle row do not interfere with each other's radial expansion, whereas there is inhibition of swarming proximal to the strain in the bottom row which is a nonswarming mutant defective in lateral flagella formation. Swarm colonies are chemotactic. Mutants with defects in *V. parahaemolyticus che* genes, although fully flagellated, fail to expand and make coherent progress on swarm plates (Sar *et al.,* 1990). Presumably strains migrate towards attractants and away from repellents. Usually swarming colonies will merge seamlessly; however occasionally strains, e.g., spontaneous variants or mutant strains, will fail to converge. In this case, sharp lines demarcating zones where growth is inhibited are formed. A similar phenomenon was observed (but not explained) many years ago with strains of *Proteus* species and a compatibility screen, the Dienes test, was used for strain typing (Dienes, 1946; Senior *et al.,* 1977). What allows some colonies to fuse and not others? Perhaps defects or differences in small-molecule signaling or reception or incompatibility of cell surface moieties leading to interference with coordination of the swarm. Chemotactic control of movement is integrated in the sense that some of the *che* genes are shared by both motility systems. It remains to be discovered if there are unique liquid- and surface-specific components to chemotaxis.

Intercellular Communication

The complex growth patterns observed for colonies on plates suggest modes of cooperative behavior. Colony dynamics must accommodate diffusion of nutrients, motility, cell division and intercellular communication (Ben-Jacob *et al.,* 1994; Shapiro, 1998). Budrene and Berg (1991) have shown complex pattern formation by motile cells of *E. coli* as gradients of attractants are established by the cells themselves. Other effects may take place at the level of gene expression. Density-dependent sensing has been postulated to be an important component of bacterial colonization and growth in communities (Batchelor *et al.,* 1997; McLean *et al.,* 1997). Biofilm architecture can be profoundly influenced by cell-to-cell signaling (Davies *et al.,* 1998). Small molecule signaling and intercellular communication may provide the cell one method for discrimination between a free-living, low cell density environment and an attached, high cell density environment (Fuqua *et al.,* 1996). *S. liquefaciens* produces two extracellular signaling molecules

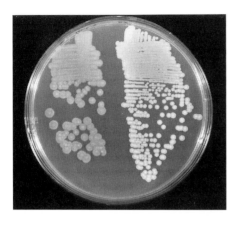

Figure 5. Opaque (OP) and Translucent (TR) Colonies are Shown on Right and Left, Respectively

belonging in the N-acyl homoserine lactone (AHL) family of autoinducers (Eberl *et al.,* 1996b). Mutants unable to produce autoinducers show defects in swarming. Experiments suggest that defects in autoinducer production lead to an inability to produce the surfactant required for swarming (Lindum *et al.,* 1998).

Much of what is known about autoinducer signaling in *V. parahaemolyticus* derives from cross-talking between *V. parahaemolyticus* and the closely related *Vibrio harveyi*. Density-dependent regulation of luminescence is well-studied in *V. harveyi* (Bassler *et al.,* 1994; Miyamoto *et al.,* 1996). The two *Vibrios* both produce two distinct autoinducer molecules, one belonging in the AHL family while the nature of the second seems unique and has not yet been determined. Although most *V. parahaemolyticus* strains are not luminescent, supernatants of cultures from these strains will induce luminescence of *V. harveyi* (Greenberg *et al.,* 1979; Bassler *et al.,* 1997). Light production in *V. harveyi* is controlled by the dual signaling systems: two autoinducers uniquely interact with cognate receptors to activate phosphorelay cascades that ultimately affect transcription of *luxR*. The product of *luxR* is a transcriptional activator of the *lux* operon, which encodes the enzymatic activities necessary for luminescence. This is an either/or system: loss of function of one signaling pathway does not eliminate luminescence or regulation of luminescence. Thus, the ultimate transcriptional activating component of the autoinducer signaling pathways in *V. harveyi* is LuxR. When the function of the gene encoding the LuxR homolog is eliminated *in V. parahaemolyticus*, swarming is unaffected (McCarter, 1998). Such mutants expand at rates comparable to the wild-type strain on rich or minimal swarm media. This suggests that autoinducer signaling mediated through LuxR is not a requirement for swarming in *V. parahaemolyticus*.

Opaque/Translucent Variation in Colony Morphology

If the LuxR homolog is not directly implicated in swarming, then what is its function in *V. parahaemolyticus*? Expression of this gene does dramatically affect other attributes of the organism. Introduction of a clone carrying the *luxR*-like locus into *V. parahaemolyticus* converts colony morphology from translucent to opaque (McCarter, 1998).

In addition to the swimmer/swarmer dimorphism, *V. parahaemolyticus* exhibits another kind of phenotypic switching. It is manifested in variable colony morphology. Descendants of a single colony can have multiple colony morphotypes. The variants are described as opaque (OP) and translucent (TR) as a result of differences in the transmission of light by the colony (Figure 5). The switching event is slow enough so that it is possible to obtain essentially uniform populations with less than 1 alternate form per 1000 cells. Properties of OP and TR are distinct and multiple traits are affected. For example, OP cells aggregate in certain kinds of liquid media, possess a thick, ruthenium-red staining capsular material, display a different array and distribution of outer membrane proteins, and swarm very poorly compared to TR cells. It is postulated that differences in cell surface characteristics lead to differential cell packing within the colony, which determines the opaque/translucent colony phenotype.

Evidence suggests that the LuxR homolog is a global regulator controlling opacity in *V. parahaemolyticus*, and so the gene has been named *opaR*. Expression of *opaR* correlates with opacity. The gene is transcribed in OP but not TR cells. When the coding sequence for *opaR* is placed under control of the pTac promoter, opacity becomes inducible by isopropyl-ß-D thiogalactopyranoside. Furthermore, disruption of the gene by transposon insertion converts an OP strain to TR. Due to the extremely high homology of the gene and its promoter region with the *V. harveyi* locus, it seems likely that *opaR* expression will be responsive to autoinducer signaling; however, this remains to be determined as does the nature of the signaling input.

DNA rearrangements occur in the *opaR* locus and this determines one basis for variation in colony morphology. Physical alterations in the DNA preclude transcription of *opaR* in the TR state. Whether the recombinational switch itself is controlled, *i.e.* responsive to environmental signals, or is a result of spontaneous genetic rearrangement is not known. So, there is potential in the system for multiple levels at which this variable phenotype can be determined. OP/TR switching alternates between expression competent and incompetent states. Transcription, in the competent state, may be regulated by environmental and or intercellular signals, and the switching event itself may be spontaneous or responsive to specific cues.

Multiple Identities of *V. parahaemolyticus*

In order to survive in changing environments, bacteria possess enormous adaptive capabilities that allow them to modulate their behavior and program gene expression in response to environmental and intercellular cues. Figure 6 illustrates potential roles for the multiple identities of *V. parahaemolyticus* and is consistent with many observations on bacterial survival in the real world (Costerton *et al.,* 1995). Motility and chemotaxis have obvious advantages for the lifestyle of the free-living swimmer cell. Moreover, swimming may allow a bacterium to find and closely approach a surface or viscous layer. Initial contact with surfaces may be facilitated by specific adhesins on the cell body or the sheathed flagellum. In liquids and/or on surfaces, cell types switch reversibly between OP and TR. Switching may occur randomly, so that a subset of the population is preadapted, or it may

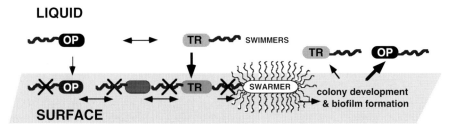

Figure 6. Multiple Identities of *V. parahaemolyticus*
Swimmer, Swarmer, Opaque (OP) and Translucent (TR) cell types allow adaptation and survival under different circumstances, for example planktonic survival in liquid versus growth in viscous environments, on surfaces, or in biofilms. OP and TR forms interconvert and may preferentially attach, or detach, to different surfaces (indicated by thick versus thin arrows). Immobilization of polar flagellar rotation (indicated by "X") signals a surface and leads to induction of surface-induced genes, including the lateral motility system and potential adhesins and virulence factors. Swarmer cell differentiation, colony development and biofilm formation is also influenced by chemical and cell-to-cell signaling.

be responsive to specific environmental conditions. The OP and TR forms, having different cell surface characteristics, may adhere preferentially to different surfaces or selectively autoaggregate and thus facilitate detachment. Once the cell makes initial contact with the surface, performance of the polar flagellum is impeded. This constitutes surface recognition, and the cell synthesizes new molecules appropriate for life on surfaces. What are these molecules? Some may aid adherence or allow protection. For other bacteria known to swarm, virulence factors are clearly produced in response to growth on surfaces (Mobley and Belas, 1995). Chemical signals, such as iron starvation, are required in addition to mechano-inactivation of flagellar rotation, to induce the swarmer cell developmental program. Differentiation to the swarmer cell allows movement over surfaces and through viscous environments. Surface-adapted cells recognize each other, and movement is coordinated and social in nature, resulting in complex multicellular behavior and growth in organized communities.

References

Allen, R., and Baumann, P. 1971. Structure and arrangement of flagella in species of the genus *Beneckea* and *Photobacterium fischeri*. J. Bacteriol. 107: 295-302.

Allison, C., Lai, H.C., Gygi, D., and Hughes, C. 1993. Cell differentiation of *Proteus mirabilis* is initiated by glutamine, a specific chemoattractant for swarming cells. Mol. Microbiol. 8: 53-60.

Atsumi, T., McCarter, L., and Imae, Y. 1992. Polar and lateral flagellar motors of marine *Vibrio* are driven by different ion membrane forces. Nature. 355: 182-184.

Atsumi, T., Maekawa, Y.,Yamada, T., Kawagishi, I., Imae, Y., and Homma, M. 1996. Effect of viscosity on swimming by the lateral and polar flagella of *Vibrio alginolyticus*. J. Bacteriol. 178: 5024-5026.

Bassler, B.L., Greenberg, E.P., and Stevens, A.M. 1997 Cross-species induction of luminescence in the quorum-sensing bacterium *Vibrio harveyi*.

J. Bacteriol. 179: 4043-4045.

Bassler, B.L., Wright, M., and Silverman, M.R. 1994. Multiple signalling systems controlling expression of luminescence in *Vibrio harveyi*: sequence and function of genes encoding a second sensory pathway. Mol. Microbiol. 13: 273-286.

Batchelor, S.E., Cooper, M., Chhabra, S.R., Glover, L.A., Stewart, G.S.A.B., Williams, P., and Prossner, J.I. 1997. Cell density-regulated recovery of starved biofilm populations of ammonia-oxidizing bacteria. Appl. Environ. Microbiol. 63: 2281-2286.

Belas, R., Simon, M., and Silverman, M. 1986. Regulation of lateral flagella gene transcription in *Vibrio parahaemolyticus*. J. Bacteriol. 167: 210-218.

Ben-Jacob, E., Schochet, O., Tenenbaum, A., Cohen, I., Czirok, A., and Vicsek, T. 1994. Generic modelling of cooperative growth patterns in bacterial colonies. Nature. 368: 46-49.

Budrene, E.O., and Berg, H.C. 1991. Complex patterns formed by motile cells of *Escherichia coli*. Nature. 349: 630-633.

CDC. 1998. Outbreak of *Vibrio parahaemolyticus* infections associated with eating raw oysters-Pacific Northwest, 1997. MMWR 47: 457-462.

Costerton, J.W., Lewandowski, Z., Caldwell, D.E., Korber, D.R., and Lappin-Scott, H.M. 1995. Microbial biofilms. Annu. Rev. Microbiol. 49: 711-745.

Davies, D.G., Parsek, M.R., Pearson, J.P., Iglewski, B.H., Costerton, J.W., and Greenberg, E.P. 1998 The involvement of cell-to-cell signals in the development of a bacterial biofilm. Science. 280: 295-8.

Dienes, L. 1946. Reproductive processes in *Proteus* cultures. Proc. Society for Experimental Biol. Med. 63: 265-270.

Eberl, L., Christiansen, G., Molin, S., and Givskov, M. 1996. Differentiation of *Serratia liquefaciens* into swarm cells is controlled by the expression of the *flhD* master operon. J. Bacteriol. 178: 554-559.

Eberl, L., Winson, M.K., Sternberg, C., Stewart, G.S.A.B., Christiansen, G., Chhabra, S.R., Bycroft, B., Williams, P., Molin, S., and Givskov, M. 1996b. Involvement of N-acyl-L-homoserine lactone autoinducers in controlling the multicellular behavior of *Serratia liquefaciens*. Mol. Microbiol. 20: 127-136.

Fuerst, J.A. 1980. Bacterial sheathed flagella and the rotary motor model for the mechanism of bacterial motility. J. Theor. Biol. 84: 761-774.

Fuqua, C., Winans, S.C., and Greenberg, E.P. 1996. Census and consensus in bacterial ecosystems: The LuxR-LuxI family of quorum sensing transcriptional regulators. Annu. Rev. Microbiol. 50: 727-751.

Furness, R.B., Fraser, G.M., Hay, N.A., and Hughes, C. 1997. Negative feedback from a *Proteus* class II flagellum export defect to the *flhDC* master operon controlling cell division and flagellum assembly. J. Bacteriol. 179: 5585-5588.

Givskov, M., Ostling, J., Eberl, L., Lindum P.W., Christiansen, G. Christiensen, A.B., Molin, S., and Kjelleberg, S. 1998. Two separate regulatory systems participate in control of swarming motility of *Serratia liquefaciens* MG1. J. Bacteriol. 180: 742-745.

Gottesman. S. 1996. Proteases and their targets in *Escherichia coli*. Annu. Rev. Genet. 30: 465-506.

Greenberg, E.P., Hastings J.W., and Ulitzur, S. 1979. Induction of luciferase synthesis in *Beneckea harveyi* by other marine bacteria. Arch. Microbiol.

120: 87-91.

Gygi, D., Rahman, M.M., Lai, H.-C., Carlson, R., Guard-Petter, J., and Hughes, C. 1995. A cell-surface polysaccharide that facilitates rapid population migration by differentiated swarm cells of *Proteus mirabilis*. Mol. Microbiol. 17: 1167-1175.

Harshey, R.M. 1994. Bees aren't the only ones: swarming in Gram-negative bacteria. Mol. Microbiol. 13: 389-394.

Harshey, R.M., and Matsuyama, T. 1994. Dimorphic transition in *Escherichia coli* and *Salmonella typhimurium*: Surface-induced differentiation into hyperflagellate swarmer cells. Proc. Natl. Acad. Sci. USA 91: 8631-8635.

Jaques, S., Kim, Y.K., and McCarter, L.L. 1999. Mutations conferring resistance to phenamil and amiloride, inhibitors of sodium-driven motility of *Vibrio parahaemolyticus*. Proc. Natl. Acad. Sci. USA. 96: 5740-5745.

Jiang, Z.Y, Rushing, B.G., Bai, Y., Gest, H., and Bauer, C.E. 1998. Isolation of *Rhodospirillum centenum* mutants defective in phototactic colony motility by transposon mutagenesis. J. Bacteriol. 180: 1248-1255.

Joseph, S.W., Colwell, R.R., and Kaper, J.B. 1982. *Vibrio parahaemolyticus* and related halophilic *Vibrios*. Crit. Rev. Microbiol. 10: 77-124.

Kaneko, T., and Colwell, R.R. 1973. Ecology of *V. parahaemolyticus* in Chesapeake Bay. J. Bacteriol. 113: 24-32.

Kaneko, T., and Colwell, R.R. 1975. Adsorption of *V. parahaemolyticus* onto chitin and copepods. Appl. Microbiol. 29: 269-274.

Kawagishi, I., Imagawa, M., Imae, Y., McCarter, L., and Homma, M. 1996. The sodium-driven polar flagellar motor of marine *Vibrio* as the mechanosensor that regulates lateral flagellar gene expression. Mol Microbiol. 20: 693-699.

Kogure, K. 1998. Bioenergetics of marine bacteria. Curr. Opin. Biotechnol. 9: 278-82.

Kogure, K., Ikemoto, E., and Morisaki, H. 1998. Attachment of *Vibrio alginolyticus* to glass surfaces is dependent on swimming speed. J. Bacteriol. 180: 932-937.

Lindum, P.W, Anthoni, U., Christophersen, C., Eberl, L., Molin, S., and Givskov, M. 1998. N-Acyl-L-homoserine lactone autoinducers control production of an extracellular lipopeptide biosurfactant required for swarming motility of *Serratia liquefaciens* MG1. J. Bacteriol. 180: 6384-6388.

Magariyama, Y., Sugiyama, S., Muramoto, K., Maekawa, Y., Kawagishi, I., Imae, Y., and Kudo, S. 1994. Very fast flagellar rotation. Nature. 371: 752.

Matsuyama, T., Kaneda, K., Nakagawa, Y., Isa, K., Hara-Hotta, H., and Yano, I. 1992. A novel extracellular cyclic lipopeptide which promotes flagellum-dependent and -independent spreading growth of *Serratia marcescens*. J. Bacteriol. 174: 1769-1776.

McCarter, L. 1994a. MotY, a component of the sodium-type flagellar motor. J. Bacteriol. 176: 4219-4225.

McCarter, L. 1994b. MotX, a channel component of the sodium-type flagellar motor. J. Bacteriol. 176: 5988-5998.

McCarter, L. 1995. Genetic and molecular characterization of the polar flagellum of *Vibrio parahaemolyticus*. J. Bacteriol. 177: 1595-1609.

McCarter, L.L. 1998. OpaR, a homolog of *Vibrio harveyi luxR*, controls opacity of *Vibrio parahaemolyticus*. J. Bacteriol. 180: 3166-3173.

McCarter, L., Hilmen, M., and Silverman, M. 1988. Flagellar dynamometer controls swarmer cell differentiation of *V. parahaemolyticus*. Cell. 54: 345-351.

McCarter, L., and Silverman, M. 1989. Iron regulation of swarmer cell differentiation of *Vibrio parahaemolyticus*. J. Bacteriol. 171: 731-736.

McCarter, L.L., and Wright, M.E. 1993. Identification of genes encoding components of the swarmer cell flagellar motor and propeller and a sigma factor controlling differentiation of *Vibrio parahaemolyticus*. J. Bacteriol. 175: 3361-3371.

McLean, R.J.C., Whitely, M., Stickler, D.J., and Fuqua, W.C. 1997. Evidence of autoinducer activity in naturally occurring biofilms. FEMS Microbiol. Lett. 154: 259-263.

Miyamoto, C.M, Chatterjee, J., Swartzman, E., Szittner, R., and Meighen, E.A. 1996. The role of the *lux* autoinducer in regulating luminescence in *Vibrio harveyi*: control of *luxR* expression. Mol. Microbiol. 19: 767-775.

Mobley, H.L., and Belas,R. 1995. Swarming and pathogenicity of *Proteus mirabilis* in the urinary tract. Trends Microbiol. 3: 280-284.

Moens, S., Schloter, M., and Vanderleyden, J. 1996. Expression of the structural gene, *laf1*, encoding the flagellin of the lateral flagella in *Azospirillum brasilense* Sp7. J. Bacteriol. 178: 5017-5019.

Ottemann, K.M., and Miller, J.F. 1997. Roles for motility in bacterial-host interactions. Mol. Microbiol. 24: 1109-1117.

O'Toole, G.A., and Kolter, R. 1998. Flagellar and twitching motility are necessary for *Pseudomonas aeruginosa* biofilm development. Mol. Microbiol. 30: 295-304.

Pratt, L.A., and Kolter, R. 1998. Genetic analysis of *Escherichia coli* biofilm formation: roles of flagella, motility, chemotaxis and type I pili. Mol. Microbiol. 30: 285-293.

Rauprich, O., Matsushita, M., Weijer, C.J., Siegert, F., Esipov, S.E., and Shapiro, J.A. 1996. Periodic phenomena in *Proteus mirabilis* swarm colony development. J. Bacteriol. 178: 6525-6538.

Sar, N., McCarter, L., Simon, M., and Silverman, M. 1990. Chemotactic control of the two flagellar systems of *Vibrio parahaemolyticus*. J. Bacteriol. 172: 334-41.

Shapiro, J.A. 1998. Thinking about bacterial populations as multicellular organisms. Annu. Rev. Microbiol. 52: 81-104.

Senior, B.W. 1977. The Dienes Phenomenon: Identification of the determinants of compatibility. J. Gen Microbiol. 102: 235-244.

Sjoblad, R.D., and Doetsch, R.N. 1982. Adsorption of polarly flagellated bacteria to surfaces. Curr. Microbiol. 7: 191-194.

Sjoblad, R.D., Emala, C.W., and Doetsch, R.N. 1983. Bacterial flagellar sheaths: structures in search of a function. Cell Motility. 3: 93-103.

Stewart, B.J., Enos-Berlage, J., and McCarter, L.L. 1997. The *lonS* gene of *Vibrio parahaemolyticus* regulates swarmer cell differentiation. J. Bacteriol. 179: 107-114.

Ulitzur, S. 1974. Induction of swarming in *V. parahaemolyticus*. Arch. Microbiol. 101: 357-363.

8

Non-Flagellar Swimming in Marine *Synechococcus*

B. Brahamsha

Marine Biology Research Division, Scripps Institution of Oceanography, University of California, San Diego, La Jolla, CA 92093-0202, USA

Abstract

Certain marine unicellular cyanobacteria of the genus *Synechococcus* exhibit a unique type of swimming motility characterized by the absence of flagella and of any other obvious organelle of motility. Although the mechanism responsible for this phenomenon remains mysterious, recent advances have included the development of testable models as well as the identification of a cell-surface polypeptide that is required for the generation of thrust. These developments, as well as the future research directions they suggest, are discussed.

Introduction

Unicellular cyanobacteria of the genus *Synechococcus* occupy an important position at the base of the marine food web. They are among the most abundant photosynthetic organisms in oligotrophic open ocean regions, where they contribute significantly to primary production (Iturriaga and Mitchell, 1986; Waterbury *et al.*, 1986; Olson *et al.*, 1990). Waterbury *et al.* (1985) reported that several isolates of marine *Synechococcus* were capable of swimming. This discovery was remarkable for two reasons. First, unlike other motile cyanobacteria which move by gliding on surfaces, motile *Synechococcus* were unique in their ability to swim in liquids. Second, marine motile *Synechococcus* had neither flagella nor any other visible organelle of locomotion. Other prokaryotes that are able to swim in liquids do so by means of flagella which function as helical propellers driven by a biological rotary motor (Hazelbauer *et al.*, 1993; Macnab, 1996). To my knowledge, the ability to swim in the absence of flagella has not been since reported for any other bacterium and appears to be confined to certain *Synechococcus* species isolated from oligotrophic marine waters (Waterbury *et al.*, 1986).

The Problem

Cells of swimming *Synechococcus* are coccoid to rod-shaped and range in size from 0.7 to 0.9 μm in diameter and 1 to 2.5 μm in length (Waterbury *et al.*, 1985). Certain features of swimming in *Synechococcus* resemble flagellar swimming. Swimming speeds range from 5 to 25 μm/sec (Waterbury *et al.*, 1985; Willey, 1988). Swimming cells rotate about their longitudinal axis as they translocate and hence generate both torque and thrust. Like flagellated bacteria (Berg and Turner, 1979), *Synechococcus* cells are slowed down and ultimately immobilized by increasing medium viscosity (Willey, 1988). Cells that become fortuitously attached to the slide or coverslip will rotate about their point of attachment at an average rate of 1 revolution/sec. Cells can rotate either in the clockwise (CW) or the counterclockwise (CCW) direction but a rotating cell has never been observed to change the direction of rotation, which, as Willey has pointed out (Willey, 1988), suggests that the "motor" rotates in only one direction and does not switch. This further implies that *Synechococcus* cells may have a "front" and a "back" end and that the direction of rotation depends on which end of the cell is tethered (Willey, 1988). Unlike flagellated bacteria, no obvious patterns in swimming behavior, such as tumbles or reversals or stops, have been observed but cell shape appears to affect motility: more coccoid cells tend to swim in more looped or spiral paths while rod-shaped cells swim in straight paths (Willey, 1988).

Initial efforts were directed at visualizing the organelle or structure responsible for the generation of torque and thrust in *Synechococcus* and included transmission electron microscopy of negatively stained whole and thin-sectioned cells (Waterbury *et al.*, 1985; Willey, 1988), and freeze-fracture and etching which has been used to visualize membrane-embedded components of the rotary motor in flagellated bacteria (Coulton and Murray, 1978; Khan *et al.*, 1988; Willey, 1988). These studies did not reveal the presence of either flagella (external or periplasmic) or other identifiable structures that might be associated with the ability to swim and in fact, the envelope structures of motile strains did not appear to differ from those of nonmotile strains (Willey, 1988). Also, an examination of actively swimming cells by high-intensity darkfield microscopy, a technique which makes possible the visualization of single flagella (Macnab, 1976), did not reveal any (Willey, 1988). Furthermore, no motility-dependent amplitude spectra were seen in images of swimming *Synechococcus* (Willey, 1988). The method used can detect the rotation of even very short cellular projections (Lowe *et al.*, 1987; Willey, 1988). Shearing experiments, which are extremely effective at eliminating motility in flagellated bacteria, did not at all affect swimming in *Synechococcus* (Waterbury *et al.*, 1985; Willey, 1988).

In flagellated bacteria, torque and thrust are generated by the rotation of the flagellum. In the absence of such appendages how then might *Synechococcus* generate these forces? As pointed out by Pitta *et al.*, (Pitta and Berg, 1995; Pitta *et al.*, 1997) this leaves the cell body itself or the cell surface as the only structures with which to generate thrust.

Models

Two models have been proposed for how *Synechococcus* might swim without flagella. One of these has been tested experimentally. Pitta and Berg (1995) ruled out self-electrophoresis as a possible propulsive mechanism by demonstrating that in a seawater-based medium, the electrophoretic mobility of *Synechococcus* was essentially 0. In such a model, a cell carries a fixed charge on its surface that is shielded by counterions in the medium. If the cell then pumps ions in at one end and pumps them out at the other end, an electric field will be set up in the external medium. In such a field, the external fluid layer containing the counterions would be driven over the cell surface like the treads of a tank and the cell would move in a direction opposite to the movement of the external fluid layer (Pitta and Berg, 1995). In a medium of high ionic strength such as seawater, a surface charge would be easily neutralized. Furthermore, because of the high conductivity of seawater, the cell would have to generate a very high current density, making self-electrophoresis for *Synechococcus* difficult, if not impossible in this environment (Pitta and Berg, 1995).

Ehlers *et al.* (1996) have proposed a model in which *Synechococcus* moves by generating traveling longitudinal or transverse surface waves, and have calculated that waves of 0.2 μm length, 0.02 μm amplitude and traveling at 160 μm/sec would allow the cell to swim at speeds of 25 μm/sec. In this model, the waves are generated by the expansion and contraction of local regions of the outer membrane and create thrust such that the cell moves in the direction of the wave. The observed rotation of the cell body during swimming can be explained by waves that are not axially symmetric. Although there is yet no experimental evidence for this model, aspects of it are certainly testable. These waves are not expected to be visible by phase contrast microscopy but Ehlers *et al.* (1996) have pointed out that it should be possible to attach particles to the cell surface and observe their behavior. The particle will vibrate at the frequency of the wave if it attaches to a region of the cell that is contracting and expanding. According to this model there is no net flow of the cell membrane and in fact, Stone and Samuel (1996) have calculated that simple surface flow is unlikely to be propulsive. There is some evidence that the outer membrane does not flow. Willey (1988) observed that beads that became attached to the cell surface served as a fixed reference point as the cell rotated, suggesting that the cell surface does not flow in bulk or move in tracks as in the gliding bacterium *Cytophaga* (Lapidus and Berg, 1982). The generation of surface waves by the contraction of helically arranged fibrils has been proposed as a mechanism for gliding in the filamentous cyanobacterium *Oscillatoria* (Halfen and Castenholz, 1971). Although this hypothesis has not been proven, the presence of such a fibrillar array between the peptidoglycan and the outer membrane of *Oscillatoria* has recently been confirmed (Adams *et al.*, 1999). Although such an array has not been observed in marine *Synechococcus*, there may be other ways of generating contractions and expansions, such as localized swelling of the cell surface (Koch, 1990; Pitta *et al.*, 1997).

Energetics

As is the case for other types of prokaryotic motility, such as flagellar swimming and gliding, the source of energy for swimming in *Synechococcus* is not ATP (Willey *et al.*, 1987). Rather, the sodium motive force is likely to be the energy source, suggesting that the device converting chemical into mechanical energy resides in the cytoplasmic membrane (Willey *et al.*, 1987). Recently, Pitta *et al.* (1997) showed that another ion, calcium, is also required for swimming. Although the role of calcium in *Synechococcus* motility is not yet understood, studies using inhibitors of voltage-gated calcium channels suggest that a calcium potential may be involved (Pitta *et al.*, 1997). These authors have suggested that calcium depolarization may serve as a way of generating a local swelling of the cell surface which could then be propagated along the cell body and hence may produce the waves postulated by Ehlers *et al.* (1996). Furthermore, calcium is required to maintain the structural integrity of the outer membrane. The specific requirements for sodium and calcium likely reflect *Synechococcus*' adaptation to its natural environment: seawater contains approximately 500 mM Na$^+$ and 10 mM Ca^{2+} and marine strains have elevated growth requirements for Na$^+$, Cl$^-$, Mg^{2+}, and Ca^{2+} (Waterbury and Rippka, 1989).

SwmA

Recently, the development of tools for the genetic manipulation of *Synechococcus* has made it possible to address the problem at the molecular level (Brahamsha, 1996b). In the course of experiments designed to detect cell-surface components that may be involved in swimming, a major polypeptide of 130 kDa was identified whose decrease in abundance following brief proteinase K treatment of actively swimming cells coincided with the loss of motility (Brahamsha, 1996a). A survey of motile and non-motile marine *Synechococcus* indicated that a polypeptide of similar abundance and molecular weight was present in all of the motile strains tested but in none of the non-motiles. This polypeptide was designated SwmA and was purified form *Synechococcus* sp. strain WH8102 and shown to be associated with the outer membrane although it does not appear to be an integral outer membrane protein (Brahamsha, 1996a). Using recently developed tools for the genetic manipulation of marine *Synechococcus* (Brahamsha, 1996b) *swmA* was insertionally inactivated in *Synechococcus* WH8102. Insertional inactivation of *swmA* results in a loss of the ability to translocate, but cells can still rotate about a point of attachment (Brahamsha, 1996a) at wild-type rates (about 1 revolution per second) (B. Brahamsha, unpublished). This suggests that SwmA functions in the generation of thrust and that thrust and torque generation are genetically separable. Furthermore, these results suggest that at least one component of the cell surface is important for this type of motility.

   SwmA is an 835-aa glycosylated polypeptide and its sequence contains a number of Ca^{2+}-binding motifs including an EF hand loop and numerous repeats of a glycine and aspartate-rich motif that has been shown to function in Ca^{2+} binding in other proteins (Baumann *et al.*, 1993). Protein similarity searches have indicated that SwmA is similar to a number of functionally diverse cell surface-associated or secreted proteins such as an S-layer protein (Gilchrist et al., 1992) and various hemolysin-like proteins (Glaser et al., 1988). However, all of these proteins also contain multiple repeats of the Ca^{2+}-binding motif and the regions of similarity to SwmA are largely confined to regions that contain the Ca^{2+} binding patterns, hence SwmA is unlikely to be functionally related to any of these (Brahamsha, 1996a). It has been suggested that calcium may play a role in the folding and stabilization of these proteins following secretion (Baumann et al., 1993). It remains to be determined whether SwmA in fact binds calcium.

SwmA is also similar to oscillin, a 646-aa Ca^{2+}-binding glycoprotein that forms an array of parallel fibrils on the surface of the gliding filamentous cyanobacterium *Phormidium uncinatum* (Hoiczyk and Baumeister, 1997), as well as to HlyA, a 322-aa partial ORF of unknown function from the non-motile filamentous cyanobacterium *Anabaena* sp. strain PCC7120 (Brahamsha, 1996a). Both oscillin and HlyA also contain multiple repeats of the Ca^{2+}-binding nonapeptide, and the regions of similarity to SwmA are for the most part confined to regions containing these repeats. These three cyanobacterial proteins have been proposed to be homologous on the basis of a stretch of similarity at their C-terminal 47 amino acids, a region that does not contain multiple repeats of the Ca^{2+}-binding motif (Hoiczyk and Baumeister, 1997). Although this is possible, this region does contain one copy of the nonapeptide within the parameters used by Hoyczyk and Baumeister (1997) (GNLYFDTNG for oscillin, GALFFDVDG for HlyA, and GVLSFDADG for SwmA).

It has been suggested that SwmA and oscillin are functionally related (Hoiczyk and Baumeister, 1997). This raises the intriguing proposition that swimming in *Synechococcus* and gliding in *Phormidium* share a common mechanism. However, although it has been proposed that oscillin is essential for gliding, this in fact has not yet been established since the non-gliding mutants that were studied lacked not only oscillin but also the S-layer underlying the oscillin fibrils and also did not produce slime (Hoiczyk and Baumeister, 1997). Oscillin is proposed to play a passive role in gliding, with its helically arranged fibrils serving to guide the secreted slime. The interaction between the extruded slime, the surface of the cell and the solid substrate would then generate thrust (Hoiczyk and Baumeister, 1997). Although this is a plausible scenario for gliding, swimming *Synechocccus* sp. do not produce extracellular slime, nor do they glide on solid surfaces (Waterbury et al., 1985; Willey, 1988). Nevertheless, perhaps SwmA alters the cell surface's characteristics or shape in such a way that rotation of the cell body results in thrust. This might occur if SwmA were distributed in a helical manner around the cell, the way oscillin fibrils are arranged, although it is difficult to predict how large these fibrils or bundles of fibrils would need to be to give the cell an "effective" helical shape such that rotation, generated by some other component, would be accompanied by forward movement. There is at least one other example in a swimming bacterium where a mutation has uncoupled cell rotation from the generation of thrust: mutants of *E. coli* with straight, rather than helical flagella, still rotate their flagella, but cannot swim (Silverman and Simon, 1974). However, no obvious differences in cell shape between the wild type and the *swmA* mutant have been detected by electron microscopy (B. Brahamsha, upublished). Polyclonal antiserum has been raised against SwmA and used in immunogold labelling of whole cells and of thin sections. The results have confirmed the biochemical evidence that SwmA is indeed localized at the cell surface and indicate that there do not appear to be localized concentrations, such as at a cell pole, for example. SwmA appears to cover the entire surface of the cell (B. Brahamsha and S. Bakalis, unpublished). However, the techniques that were used to visualize SwmA may not have sufficiently preserved possible substructures and the use of cryotechniques such as the ones used to visualize oscillin (Hoiczyk and Baumeister, 1995; Hoiczyk and Baumeister, 1997) may be required. Efforts are under way to determine more precisely SwmA's arrangement on the cell surface (B. Brahamsha, unpublished).

Alternatively, could SwmA be involved in a contractile mechanism that would generate the traveling waves proposed by Ehlers *et al.* (1996)? The fact that SwmA mutants still rotate is not necessarily inconsistent with this model. SwmA could affect certain properties of the waves, such as their amplitude for example. Could SwmA's putative Ca^{2+}-binding sites be involved in a contractile mechanism? Calcium is required for swimming in *Synechocccus* (Pitta *et al.*, 1997) and the calcium analog terbium inhibits motility in the presence of calcium in a competitive fashion. Swimming speeds decrease linearly with increasing terbium which led Pitta *et al.* (1997) to suggest that there may be multiple force generators that act independently and that are inhibited in an all or none fashion. Does SwmA bind terbium? It is of course possible that SwmA physically interacts with another protein or proteins to generate thrust. Efforts are underway to identify proteins which might interact with SwmA (B. Brahamsha, unpublished).

If swimming in *Synechococcus* is anywhere as complex as two other types of prokaryotic motility, flagellar swimming and gliding, many genes and gene products will be involved. For instance, in *E. coli*, in addition to genes whose products are required for chemotaxis, there are over 40 genes involved in the synthesis and assembly of a functional flagellum (Macnab, 1996). In *M. xanthus*, about 70 loci have been identified which control the gliding movement of individual cells and groups of cells (Hartzell and Youderian, 1995). SwmA is clearly not the only protein involved in swimming. For example, what is the torque generator and what are its components? What are the components involved in transducing the sodium motive force into mechanical energy? Mutant screens are currently underway to identify some of these players. (B. Brahamsha, unpublished). A gene has been identified which is required for the expression of SwmA and may be involved in its glycosylation (B. Brahamsha, in preparation).

Ecological Implications

Why do some marine *Synechococcus* swim? Unlike some gliding cyanobacteria, marine *Synechococcus* appear to exhibit neither phototactic nor photophobic responses (Willey and Waterbury, 1989). Willey and Waterbury (1989) calculated that a cell swimming at 25 μm/sec in a straight line would cover a distance of only 2 m in 24 hrs. Such a distance in the euphotic zone of oligotrophic waters would result in negligible differences in either light quantity or quality and hence phototactic or phobic responses would confer no advantage. Motile strains of marine *Synechococcus* are not usually found in nutrient-rich coastal regions but rather in oligotrophic regions of the world's oceans where nitrogen is often limiting (Waterbury *et al.*, 1986). Willey and Waterbury (1989) have shown that at least one motile strain of *Synechococcus* (WH8113) is chemotactic toward nitrogenous compounds such as ammonia, nitrate, ß-alanine, glycine, and urea at ecologically relevant concentrations. They have suggested that for organisms living in oligotrophic environments, the ability to detect and seek out patches of nutrient enrichment, such as larger phytoplankton cells or microaggregates (marine snow), may represent an ecological advantage. Indeed, such marine microaggregates are heavily colonized by bacteria and have been shown to be the site of a number of hydrolytic activities, including proteolysis (Smith *et al.*, 1992). It is possible that the resulting surrounding gradients of nitrogenous compounds are detected by motile *Synechococcus* species. Recent work has shown that such patches could serve to attract swarms of bacteria (Blackburn *et al.*, 1998). How cells of *Synechococcus* might modulate their swimming behavior to respond to chemical gradients remains to be determined. Unlike flagellated swimming bacteria, no behaviors such as tumbles or reversals have been observed in marine *Synechococcus*.

This brief review has raised more questions than it has answered, not the least of which is still how does *Synechococcus* manage to swim without flagella. It is hoped that the use of molecular tools to identify components of the motility machinery coupled with the testing of proposed models will yield further insights into this fascinating problem.

Acknowledgements

I thank Tom Pitta, Howard Berg and John Waterbury for stimulating discussions. This work was supported by a grant from the NSF to BB (MCB-9727759).

References

Adams, D.G., Ashworth, D., and Nelmes, B. 1999. Fibrillar array in the cell wall of a gliding filamentous cyanobacterium. J. Bacteriol. 181: 884-892.

Baumann, U., Wu, S., Flaherty, K.M., and McKay, D.B. 1993. Three-dimensional structure of the alkaline protease of *Pseudomonas aeruginosa*: a two-domain protein with a calcium binding parallel beta roll motif. EMBO J. 12: 3357-3364.

Berg, H.C., and Turner, L. 1979. Movement of microorganisms in viscous environments. Nature. 278: 349-351.

Blackburn, N., Fenchel, T., and Mitchell, J. 1998. Microscale nutrient patches in planktonic habitats shown by chemotactic bacteria. Science. 282: 2254-2256.

Brahamsha, B. 1996a. An abundant cell-surface polypeptide is required for swimming by the nonflagellated marine cyanobacterium *Synechococcus*. Proc. Natl. Acad. Sci. USA. 93: 6504-6509.

Brahamsha, B. 1996b. A genetic manipulation system for oceanic cyanobacteria of the genus *Synechococcus*. Appl. Environ. Microbiol. 62: 1747-1751.

Coulton, J.W., and Murray, R.G.E. 1978. Cell envelope associations of *Aquaspirillum serpens* flagella. J. Bacteriol. 136: 1037-1049.

Ehlers, K.M., Samuel, A.D.T., Berg, H.C., and Montgomery, R. 1996. Do cyanobacteria swim using traveling surface waves? Proc. Natl. Acad. Sci. USA. 93: 8340-8343.

Gilchrist, A., Fisher, J.A., and Smit, J. 1992. Nucleotide sequence analysis of the gene encoding the *Caulobacter crescentus* paracrystalline surface layer protein. Can. J. Microbiol. 38: 193-202.

Glaser, P., Ladant, D., Sezer, O., Pichot, F., Ullmann, A., and Danchin, A. 1988. The calmodulin-sensitive adenylate cyclase of *Bordetella pertussis*: cloning and expression in *E. coli*. Mol. Microbiol. 2: 19-30.

Halfen, L.N., and Castenholz, R.W. 1971. Gliding in a blue-green alga, *Oscillatoria princeps*. J. Phycol. 7: 133-145.

Hartzell, P.L., and Youderian, P. 1995. Genetics of gliding motility and development in *Myxococcus xanthus*. Arch. Microbiol. 164: 309-323.

Hazelbauer, G.L., Berg, H.C., and Matsumura, P. 1993. Bacterial motility and signal transduction. Cell. 73: 15-22.

Hoiczyk, E., and Baumeister, W. 1995. Envelope structure of four gliding filamentous cyanobacteria. J. Bacteriol. 177: 2387-2395.

Hoiczyk, E., and Baumeister, W. 1997. Oscillin, an extracellular, Ca^{2+}-binding glycoprotein essential for the gliding motility of cyanobacteria. Mol. Microbiol. 26: 699-708.

Iturriaga, R., and Mitchell, B.G. 1986. Chroococcoid cyanobacteria: a significant component in the food web dynamics of the open ocean. Mar. Ecol. Prog. Ser. 28: 291-297.

Khan, S., Dapice, M., and Reese, T.S. 1988. Effects of *mot* gene expression on the structure of the flagellar motor. J. Mol. Biol. 202: 575-584.

Koch, A.L. 1990. The sacculus contraction/expansion model for gliding motility. J. Theoret. Biol. 142: 95-112.

Lapidus, I.R., and Berg, H.C. 1982. Gliding motility of *Cytophaga* sp. strain U67. J. Bacteriol. 151: 384-398:

Lowe, G.E., Meister, M., and Berg, H.C. 1987. Rapid rotation of flagellar bundles in swimming bacteria. Nature. 325: 637-640.

Macnab, R.M. 1976. Examination of bacterial flagellation by dark field microscopy. J. Clin. Microbiol. 4: 258-265.

Macnab, R.M. 1996. Flagella and motility. In: *Escherichia coli* and *Salmonella*: Cellular and Molecular Biology. F.C. neidhardt, ed. ASM Press, Washington D. C. p. 123-145.

Olson, R.J., Chisholm, S.W., Zettler, E.R., and Armbrust, E.V. 1990. Pigments, size, and distribution of *Synechococcus* in the North Atlantic and Pacific Oceans. Limnol. Oceanog. 35: 45-58.

Pitta, T.P., and Berg, H.C. 1995. Self-electrophoresis is not the mechanism for motility in swimming cyanobacteria. J. Bacteriol. 177: 5701-5703.

Pitta, T.P., Sherwood, E.E., Kobel, A.M., and Berg, H.C. 1997. Calcium is required for swimming by the nonflagellated cyanobacterium *Synechococcus* strain WH8113. J. Bacteriol. 179: 2524-2528.

Silverman, M., and Simon, M. 1974. Flagellar rotation and the mechanism of bacterial motility. Science. 249: 73-74.

Smith, D.C., Simon, M., Alldredge, A.L., and Azam, F. 1992. Intense hydrolytic activity on marine aggregates and implications for rapid particle dissolution. Nature. 359: 139-142.

Stone, H.A., and Samuel, A.D.T. 1996. Propulsion of microorganisms by surface distortions. Physical Rev. Lett. 77: 4102-4104.

Waterbury, J.B., and Rippka. 1989. Order Chroococcales Wettstein 1924, emend. Rippka *et al.*, 1979. Bergey's Manual of Systematic Bacteriology. Williams and Wilkins, Baltimore. p. 1728-1746.

Waterbury, J.B., Watson, S.W., Valois, F.W., and Franks, D.G. 1986. Biological and ecological characterization of the marine unicellular cyanobacterium *Synechococcus*. Can. Bull. Fish. Aquat. Sci. 214: 71-120.

Waterbury, J.B., Willey, J.M., Franks, D.G., Valois, F.W., and Watson, S.W. 1985. A cyanobacterium capable of swimming motility. Science. 230: 74-76.

Willey, J.M. 1988. Characterization of swimming motility in a marine unicellular cyanobacterium. Ph.D. thesis. Woods Hole Oceanographic institution and Massachussetts Institute of Technology, Cambridge, MA.

Willey, J.M., and Waterbury, J.B. 1989. Chemotaxis toward nitrogenous compounds by swimming strains of marine *Synechococcus* spp. App. Environ. Microbiol. 55: 1888-1894.

Willey, J.M., Waterbury, J.B., and Greenberg, E.P. 1987. Sodium-coupled motility in a swimming cyanobacterium. J. Bacteriol. 169: 3429-3434.

9

Petroleum Biodegradation in Marine Environments

Shigeaki Harayama, Hideo Kishira, Yuki Kasai and Kazuaki Shutsubo

Marine Biotechnology Institute
3-75-1 Heita, Kamaishi, Iwate 026-0001, Japan

Abstract

Petroleum-based products are the major source of energy for industry and daily life. Petroleum is also the raw material for many chemical products such as plastics, paints, and cosmetics. The transport of petroleum across the world is frequent, and the amounts of petroleum stocks in developed countries are enormous. Consequently, the potential for oil spills is significant, and research on the fate of petroleum in a marine environment is important to evaluate the environmental threat of oil spills, and to develop biotechnology to cope with them.

Crude oil is constituted from thousands of components which are separated into saturates, aromatics, resins and asphaltenes. Upon discharge into the sea, crude oil is subjected to weathering, the process caused by the combined effects of physical, chemical and biological modification. Saturates, especially those of smaller molecular weight, are readily biodegraded in marine environments. Aromatics with one, two or three aromatic rings are also efficiently biodegraded; however, those with four or more aromatic ring are quite resistant to biodegradation. The asphaltene and resin fractions contain higher molecular weight compounds whose chemical structures have not yet been resolved. The biodegradability of these compounds is not yet known.

It is known that the concentrations of available nitrogen and phosphorus in seawater limit the growth and activities of hydrocarbon-degrading microorganisms in a marine environment. In other words, the addition of nitrogen and phosphorus fertilizers to an oil-contaminated marine environment can stimulate the biodegradation of spilled oil. This notion was confirmed in the large-scale operation for bioremediation after the oil spill from the *Exxon Valdez* in Alaska.

Many microorganisms capable of degrading petroleum components have been isolated. However, few of them seem to be important for petroleum biodegradation in natural environments. One group of bacteria belonging to the genus *Alcanivorax* does become predominant in an oil-contaminated marine environment, especially when nitrogen and phosphorus fertilizers are added to stimulate the growth of endogenous microorganisms.

Introduction

Petroleum is a viscous liquid mixture that contains thousands of compounds mainly consisting of carbon and hydrogen. Oil fields are not uniformly distributed around the globe, but being in limited areas such as the Persian Gulf region. The world production of crude oil is more than three billion tons per year, and about the half of this is transported by sea. Consequently, the international transport of petroleum by tankers is frequent. All tankers take on ballast water which contaminates the marine environment when it is subsequently discharged. More importantly, tanker accidents exemplified by that of the T/V *Exxon Valdez* in Prince William Sound, Alaska, severely affect the local marine environment. Off-shore drilling is now common to explore new oil resources and this constitutes another source of petroleum pollution. However, the largest source of marine contamination by petroleum seems to be the runoff from land. Annually, more than two million tons of petroleum are estimated to end up in the sea. Fortunately, petroleum introduced to the sea seems to be degraded either biologically or abiotically (Readman *et al.*, 1992). In this article, the fate of petroleum in a marine environment is reviewed, with special emphasis placed on its biodegradation.

Components of Petroleum

All petroleum products are derived from crude oil whose major constituents are hydrocarbons. Petroleum components can be separated into four fractions, the saturated, aromatic, resin and asphaltene fractions, by absorption chromatography. Each of these fractions contains a large number of compounds (Karlsen and Larter, 1991).

Saturates are hydrocarbons containing no double bonds. They are further classified according to their chemical structures into alkanes (paraffins) and cycloalkanes (naphthenes). Alkanes have either a branched or unbranched (normal) carbon chain(s), and have the general formula C_nH_{2n+2}. Cycloalkanes have one or more rings of carbon atoms (mainly cyclopentanes and cyclohexanes), and have the general formula C_nH_{2n}: The majority of cycloalkanes in crude oil have an alkyl substituent(s) (Figure 1). Aromatics have one or more aromatic rings with or without an alkyl substituent(s). Benzene is the simplest one (Figure 1), but alkyl-substituted aromatics generally exceed the non-substituted types in crude oil (Mater and Hatch, 1994). In contrast to the saturated and aromatic fractions, both the resin and asphaltene fractions contain non-hydrocarbon polar compounds. Their elements contain, in addition to carbon and hydrogen, trace amounts of nitrogen, sulfur and/or oxygen. These compounds often form complexes

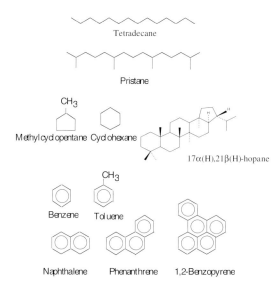

Figure 1. Representative Hydrocarbons
Tetradecane (an *n*-alkane), pristane (a branched alkane), and methylcyclopentane, cyclohexane and 17α[H],21ß[H]-hopane (cycloalkane compounds) are present in the saturated fraction of crude oil. The other compounds shown in this figure are present in the aromatic fraction.

with heavy metals. Asphaltenes consist of high-molecular-weight compounds which are not soluble in a solvent such as *n*-heptane, while resins are *n*-heptane-soluble polar molecules. Resins contain heterocyclic compounds, acids and sulfoxides.

The components of petroleum in crude oil have been analyzed mainly by using gas chromatography in combination with mass spectrometry (GC/MS). Consequently, the chemical structures of the higher-molecular-weight components (the heavy fractions) that cannot be identified by GC are mostly unknown. Furthermore, the compositions of many branched alkanes and alkyl cyclo-alkanes have not been determined because their isomers are numerous and cannot be resolved by GC (Killops and Al-Juboori, 1990; Gough and Rowland, 1990). Therefore, a multitude of analytical techniques such as flame ionization detection, IR- and UV-absorption spectrometry, NMR and elemental analysis in combination with appropriate separation techniques such as various chromatographic methods and/or chemical conversion is necessary to characterize petroleum, and especially its heavy fractions.

Various petroleum products are produced by refining crude oil. Refining is essentially a fractional distillation process by which different fractions or cuts are produced. Alkenes, a series of unsaturated hydrocarbons including ethylene, are not found in crude oil, but are produced during the cracking of crude oil.

Behavior of Petroleum in Marine Environments

When petroleum is spilled into the sea, it spreads over the surface of the water. It is subjected to many modifications, and the composition of the petroleum changes with time. This process is called weathering, and is mainly due to evaporation of the low-molecular-weight fractions, dissolution of the water-soluble components, mixing of the oil droplets with seawater, photochemical oxidation, and biodegradation.

Those petroleum components with a boiling point below 250 °C are subjected to evaporation. Therefore, the content of *n*-alkanes, whose chain length is shorter than C14, is reduced by weathering. The content of aromatic hydrocarbons within the same boiling point range is also reduced as they are subjected to both evaporation and dissolution. The mixing of oil with seawater occurs in several forms. Dispersion of the oil droplets into a water column is induced by the action of waves, while water-in-oil emulsification occurs when the petroleum contains polar components that act as emulsifiers. A water-in-oil emulsion containing more than 70% of seawater becomes quite viscous; it is called chocolate mousse from its appearance. After the light fractions have evaporated, heavy residues of petroleum can aggregate to form tar balls whose diameter ranges from microscopic size to several tenths of a centimeter.

After a large oil spill, the oil slick is sometimes treated with a dispersant. Dispersants emulsify petroleum by reducing the interfacial tension between petroleum and water. The small droplets that are formed are dispersed into a water column to a depth of several meters, preventing wind-induced drift of the oil slick. It is claimed that treatment by a dispersant enhances the biodegradation of petroleum. However, the results of such tests are controversial (Tjessem *et al.*, 1984). The original dispersants used were highly toxic; however, less toxic dispersants have subsequently been developed.

Under sunlight, petroleum discharged at sea is subjected to photochemical modification. Some reports have suggested the light-induced polymerization of petroleum components, while others have suggested their photodegradation. An increase in the polar fraction and a decrease in the aromatic fraction have also been observed. Aliphatic components do not significantly absorb solar light, and are by themselves photochemically inert. However, they can be degraded by photosensitized oxidation. The aromatic or polar components in petroleum and anthraquinone that is present in seawater can provoke the degradation of *n*-alkanes into terminal *n*-alkenes (a carbon-carbon double bond at position 1) and low-molecular-weight carbonyl compounds (Ehrhardt and Weber, 1991).

The water-soluble components of petroleum exert a toxic effect on marine organisms. In general, aromatic compounds are more toxic than aliphatic compounds, and smaller molecules are more toxic than larger ones in the same series. Solar irradiation affects oil toxicity: Surface films become less toxic due to the loss of polycyclic aromatic hydrocarbons, but the toxicity of the water-soluble fraction increases as its concentration increases (Nicodem *et al.*, 1997).

Biodegradation of Petroleum Components

n-Alkanes, a major group in crude oil, are readily biodegraded in the marine environment. *n*-Alkanes are aerobically biodegraded by several pathways. The degradation of *n*-alkanes of medium chain length by *Pseudomonas putida* containing the OCT plasmid is initiated by alkane hydroxylase. This enzyme consists of three components: the membrane-bound oxygenase component, and two soluble components called rubredoxin and rubredoxin reductase. The catalytic center of the oxygenase component contains a dinuclear iron cluster which is also found in other enzymes such as methane monooxygenase and ribonucleotide reductase (Shanklin *et al.*, 1997). In *P. putida* (OCT), oxidation of the methyl group of *n*-alkanes by alkane hydroxylase yields *n*-alcanols that are further oxidized by a membrane-bound alcohol dehydrogenase to *n*-alkanals. The *n*-alcanals are subsequently transformed to fatty acids and then to acyl CoA by aldehyde dehydrogenase and acyl-CoA synthetase, respectively (Figure 2; van Beilen *et al.*, 1994). An *n*-alkane-degrading pathway yielding secondary alcohols has also been reported. In this pathway, *n*-alkanes are oxidized by monooxygenase to secondary alcohols, then to ketones, and finally to fatty acids (Figure 2; Markovetz and Kallio, 1971; Whyte *et al.*, 1998). In *Acinetobacter* strain M-1, *n*-alkanes are transformed to *n*-alkyl peroxides, and these molecules would be further metabolized to the corresponding aldehyde. The first enzyme involved in this pathway contains FAD^+ and Cu^{2+} as prosthetic groups (Figure 2; Maeng *et al.*, 1996).

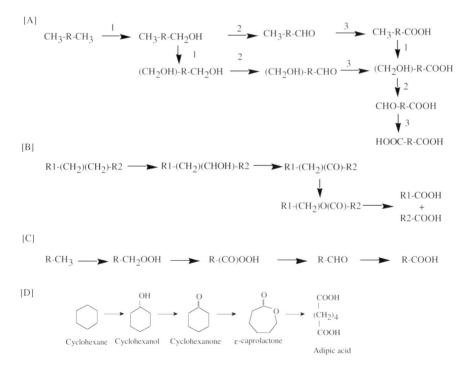

Figure 2. Alkane Degradative Pathways
[A] Terminal oxidation of *n*-alkanes. α- and ω-hydroxylation is catalyzed by the same set of enzymes. With bacteria, steps 1, 2 and 3 are catalyzed by alkane monooxygenase, fatty alcohol dehydrogenase and fatty aldehyde dehydrogenase, respectively. With yeast, step 1 is catalyzed by P450 monooxygenase, while steps 2 and 3 are catalyzed either by fatty alcohol oxidase and fatty aldehyde dehydrogenase, respectively, or by the P450 monooxygenase involved in step 1; [B] Subterminal oxidation of *n*-alkanes; [C] *n*-Alkane degradation via alkyl hydroperoxides; [D] Degradation of cyclohexane.

In the 1950s and 1960s, the microbial degradation of petroleum attracted particular attention as petroleum hydrocarbons are considered to be non-expensive substrates for producing biomass. Many yeast species, e.g. *Candida maltosa*, *Candida tropicalis* and *Candida apicola*, were investigated for use with *n*-alkanes. The first step of alkane degradation (terminal hydroxylation) and of ω-hydroxylation is catalyzed by P450 monooxygenase. The alcohols thus formed are processed by fatty alcohol oxidase and fatty aldehyde dehydrogenase. The P450 enzyme from some yeast strains can catalyze not only the terminal hydroxylation of long-chain *n*-alkanes and the ω-hydroxylation of fatty acids, but also the subsequent two steps to yield fatty acids and α,ω-dioic acids (Figure 2; Scheller *et al.*, 1998). Catabolic pathways for the degradation of branched alkanes have been elucidated for a few bacteria; for example, *Rhodococcus* strain BPM 1613 degraded phytane (2,6,10,14-tetramethylhexadecane), norpristane (2,6,10-trimethylpentadecane) and farnesane (2,6,10-trimethyldodecane) via ß-oxidation (Nakajima *et al.*, 1985).

Cycloalkanes, including condensed cycloalkanes are degraded by a co-oxidation mechanism. The formation of a cyclic alcohol and a ketone has been observed. A monooxygenase introduces an oxygen into the cyclic ketone, and the cyclic ring is cleaved (Figure 2). The degradation of substituted cycloalkanes seems to be less difficult than that of unsubstituted cycloalkanes (Morgan and Watkinson, 1994).

A multitude of catabolic pathways for the degradation of aromatic compounds have been elucidated; for example, toluene is degraded by bacteria along five different pathways. On the pathway encoded by the TOL plasmid, toluene is successively degraded to benzyl alcohol, benzaldehyde and benzoate, which is further transformed to the TCA cycle intermediates. The first step of toluene degradation with *P. putida* F1 is the introduction of two hydroxyl groups to toluene, forming *cis*-toluene dihydrodiol. This intermediate is then converted to 3-methylcatechol. With *Pseudomonas mendocina* KR1, toluene is converted by toluene 4-monooxygenase to *p*-cresol, this being followed by *p*-hydroxybenzoate formation through oxidation of the methyl side chain. With *Pseudomonas pickettii* PKO1, toluene is oxidized by toluene 3-monooxygease to *m*-cresol, which is further oxidized to 3-methylcatechol by another monooxygenase. With *Bukholderia cepacia* G4, toluene is metabolized to *o*-cresol by toluene 2-monooxygenase, this intermediate being transformed by another monooxygenase to 3-methylcatechol. *Burkholderia* sp. strain JS150 is unique in using multiple pathways for the metabolism of toluene (Figure 3; Johnson and Olsen, 1997).

Simple polynuclear aromatic hydrocarbons (PAHs) such as naphthalene, biphenyl and phenanthrene are readily degraded aerobically. The degradation of these compounds is generally initiated by dihydroxylation of one of the polynuclear aromatic rings, this being followed by cleavage of the dihydroxylated ring. Ring hydroxylation is catalyzed by a multi-component dioxygenase which consists of a reductase, a ferredoxin, and an iron sulfur protein, while ring cleavage is generally catalyzed by a iron-containing *meta*-cleavage enzyme. The carbon skeleton produced by the ring-cleavage reaction is then dismantled, before cleavage of the second aromatic ring (Saito *et al.*, 1999; Harayama *et al.*, 1992; Figure 4).

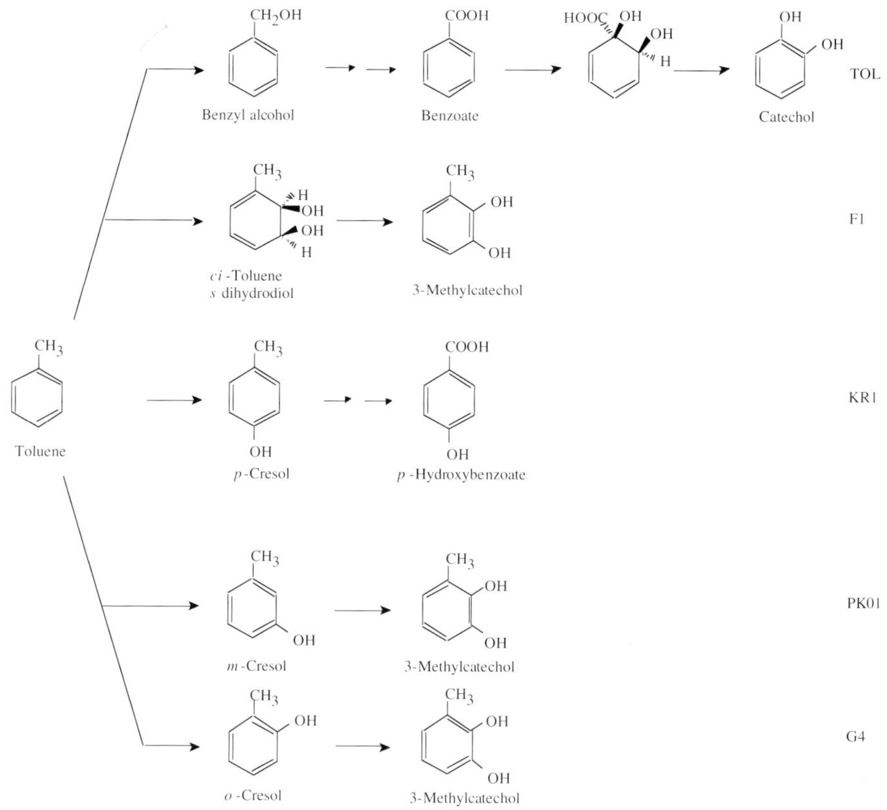

Figure 3. Divergent Pathways for the Aerobic Degradation of Toluene
The pathways from the top to the bottom are found with *P. putida* (TOL), *P. putida* F1, *P. mendocina* KR1, *P. pickettii* PKO1, and *B. cepacia* G4, respectively.

PAHs possessing four or more fused aromatic rings have very low water solubility and tend to be adsorbed to a solid surface. These characteristics constitute a major constraint for biodegradation. *Mycobacteria* and *Sphingomonas* have been isolated as bacteria that are able to degrade PAHs possessing four or more fused aromatic rings. Some *mycobacteria* mineralize (degrade into CO_2 and H_2O) pyrene, fluoranthene and benzo[*a*]pyrene (Harayama, 1997).

Asphaltenes and resins are considered to be recalcitrant to biodegradation. Rontani *et al.* (1985) have reported that about half of the asphaltene components were co-metabolically oxidized by a microbial consortium in the presence of C12 to C18 *n*-alkanes. These observations should be reinvestigated as the preparation of pure fractions of resins and asphaltenes is difficult and these fractions almost always contain saturates and aromatics (Myhr *et al.*, 1989).

Petroleum components which are trapped in marine sediment tend to persist under anaerobic conditions. Nevertheless, ecological studies have demonstrated that certain hydrocarbons can be oxidized under anaerobic

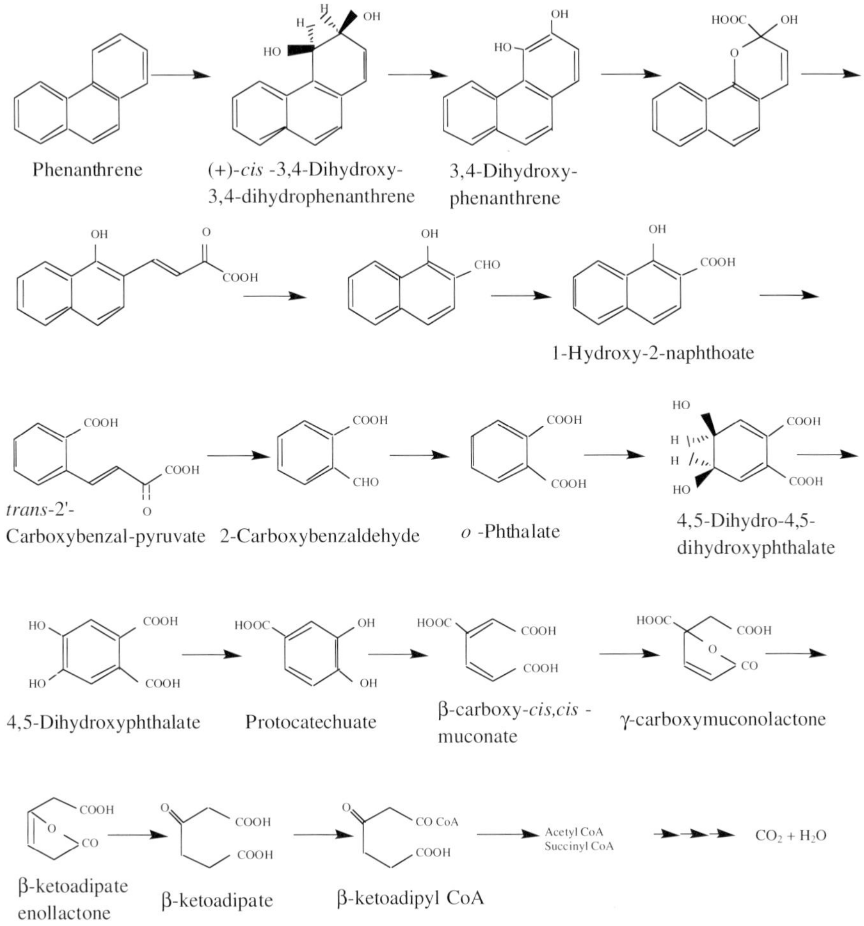

Phenanthrene

(+)-*cis* -3,4-Dihydroxy-
3,4-dihydrophenanthrene

3,4-Dihydroxy-
phenanthrene

1-Hydroxy-2-naphthoate

trans-2'-
Carboxybenzal-pyruvate   2-Carboxybenzaldehyde   *o* -Phthalate

4,5-Dihydro-4,5-
dihydroxyphthalate

4,5-Dihydroxyphthalate   Protocatechuate   β-carboxy-*cis,cis* -
muconate

γ-carboxymuconolactone

β-ketoadipate
enollactone

β-ketoadipate

β-ketoadipyl CoA

Acetyl CoA
Succinyl CoA

$CO_2 + H_2O$

Figure 4. Catabolic Pathway for the Degradation of Phenanthrene with *Nocardioides* sp. KP7

conditions when either nitrate reduction, sulfate reduction, methane production, Fe(III) reduction or photosynthesis is coupled to the hydrocarbon oxidation (Evans and Fuchs, 1988). Many hydrocarbons, such as alkanes, alkenes and aromatic hydrocarbons including benzene, toluene, xylenes, ethyl- and propylbenzenes, trimethylbenzenes, naphthalene, phenanthrene and acenaphthene, are known to be anaerobically degraded (Bregnard *et al.*, 1997). The pathways for the degradation of alkanes and alkenes are not yet clear. Anaerobic bacterium strain HD-1 grows on CO_2 in the presence of H_2 or tetradecane. In the absence of H_2, tetradecane is degraded, and the major metabolic intermediate is 1-dodecene (Morikawa *et al.* 1996).

Many pathways for the anaerobic degradation of toluene have recently been proposed. All of these pathways transform the initial substrate into the common intermediate, benzoyl-coenzyme A (CoA). With *Thauera* sp. strain T1, the oxidation of toluene is initiated by the formation of benzylsuccinate from toluene and fumarate. The genes for the pathway enzymes and the regulatory proteins have been characterized (Coschigano *et al.*, 1998). After the formation of benzyl-CoA, this intermediated is reduced by benzoyl-CoA reductase to yield cyclohex-1,5-diene-1-carboxyl-CoA. The subsequent steps, however, are controversial. With *Rhodospeudomonas palustris*, the diene intermediate is reduced to cyclohex-1-ene-1-carboxyl-CoA, while with *Thauera aromatica*, it is hydrated to 6-hydroxycyclohex-1-ene-1-carboxyl-CoA (Figure 5; Egland *et al.*, 1997; Breese *et al.*, 1998).

Biodegradation of Emulsified Petroleum

Most of the petroleum-degrading bacteria (bacteria capable of growing on Arabian crude oil as the sole source of carbon and energy) produce surfactants or emulsifiers (our unpublished observations). Therefore, cultures of these bacteria become brown and turbid as an oil slick is transformed into many small oil droplets. Bacterial cells are associated on the surface of the droplets, and such contact may facilitate the assimilation of petroleum components into the cells. This observation suggests that surfactants/ emulsifiers promote the biodegradation of petroleum components. However, conflicting results have been reported concerning the effect of surfactants/ emulsifiers on the biodegradation of hydrocarbons (Rosenberg and Ron, 1997). Emulsan produced by *Acinetobacter calcoaceticus* RAG-1 inhibited the biodegradation of crude oil (Foght *et al.*, 1989) while rhamnolipide produced by *Pseudomonas aeruginosa* enhanced the biodegradation of octadecane (Zhang and Miller, 1992).

Petroleum Biodegradation in Marine Environments

Many catastrophic oil spills from large tanker accidents have attracted public attention to the fate of petroleum hydrocarbons in marine environments. In response to this concern, research into the biodegradation of petroleum in

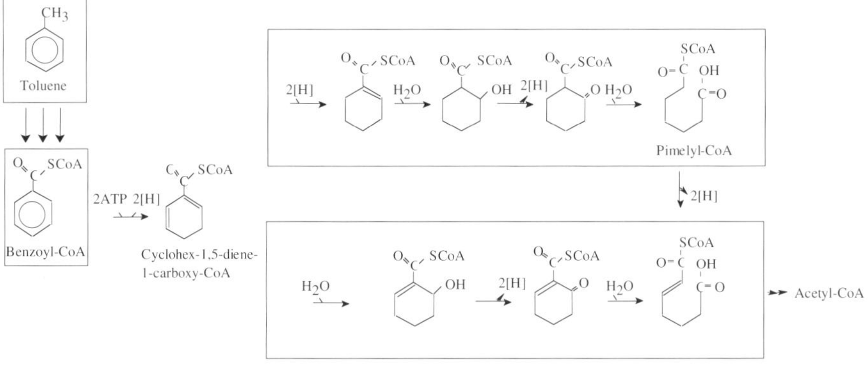

Figure 5. Anaerobic Degradation of Toluene
Several routes are proposed for the transformation of toluene to benzoyl-CoA. After the conversion of benzoyl-CoA into cyclohex-1-diene-1-carboxyl-CoA, this product is processed differently with two different bacteria, *R. palustris* (upper box) and *Thauera aromatica*. (lower box).

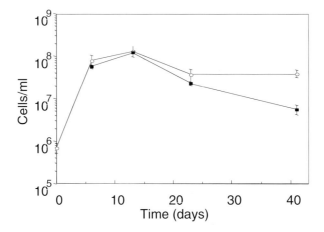

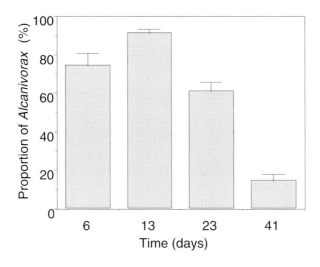

Figure 6. Population Analysis of Crude-oil Degrading Microorganisms
Natural seawater was sampled from Kamaishi Bay (Japan) to which NH_4NO_3 (1 g/l), K_2HPO_4 (0.2 g/l), ferric citrate (20 mg/l) and Arabian light crude oil (2 g/l) were supplemented. The culture was grown at 20 °C while shaking.[A] Time-course plot of the total number of bacteria determined by DAPI staining and that of the *Alcanivorax* population determined by the FISH method. [B] Time-course plot of the *Alcanivorax* population ratio.

natural environments has been intensified. The pioneering studies by Atlas, Bartha and their colleagues (Atlas and Bartha, 1993; Prince, 1993) have demonstrated that the available concentrations of nitrogen and phosphorus in seawater are limiting factors for the growth of hydrocarbon-degrading microorganisms. Thus, the addition of nitrogen and phophorus fertilizers stimulates the biodegradation of petroleum. In general, small hydrocarbon molecules are more easily biodegraded than larger ones, and aromatics are degraded at a much slower rate than that of alkanes in marine environments (Oudot, 1984; Kennicut, 1988; Ishihara et al., 1995; Sugiura et al., 1997; Sasaki et al., 1998; Wang et al., 1998).

Many oil spills in the sea cause shoreline pollution, despite efforts to prevent the drift of a spill toward the coastline. Cleaning up a polluted coastline by enhancing microbial activities was first attempted in 1989 after the spill from the Exxon Valdez. The initial measure taken after this accident was physical washing with high-pressure water. Subsequently, fertilizers were applied to the polluted beaches to accelerate the growth and activities of petroleum-degrading microorganisms. Two to three weeks later, pebbles on the beaches that had been treated with fertilizers had become significantly cleaner than those in the control area (Pritchard and Costa, 1991). Nevertheless, it was difficult to evaluate the effect of the treatment due to heterogeneity in the oil contamination of samples. Five-ring cycloalkanes, hopanes (Figure 1), are frequently used as a conserved quantitative internal standard because they are resistant to biodegradation. Using hopanes as an internal standard, it has been demonstrated that the fertilizer application significantly increased the rate of petroleum biodegradation, and that about half of the petroleum had been removed within three months after applying fertilizers in sufficient amounts (Bragg et al., 1994).

Although hydrocarbons have usually been found to persist under strict anaerobic conditions, these compounds are degraded in some types of marine harbor sediment under sulfate-reducing conditions. When hydrocarbon-degrading sediment is used to inoculate a type of sediment that shows no hydrocarbon-degrading activity under anaerobic conditions, such activity is generated. This observation indicates that hydrocarbon contamination could be treated under sulfate-reducing conditions, and that the seeding (introduction of foreign microorganisms or equivalent samples) of anaerobic microbial consortia adapted to specific hydrocarbons would be effective to enhance the anaerobic biodegradation of these hydrocarbons (Coates et al., 1997; Weiner and Lovley, 1998).

Alcanivorax as a Group of Bacteria Predominant in Oil-Contaminated Seawater

Hydrocarbon-degrading bacteria and fungi are widely distributed in marine habitats (Floogate, 1984). However, their roles in a natural marine environment or in oil-contaminated soil with or without treatment by bioremediation (fertilizer application to enhance microbial activities) are largely unknown. Our recent research has made it clear that bacteria belonging to the genus Alcanivorax are important for the biodegradation of petroleum, especially under bioremediation conditions.

The Alkanivorax group was first isolated from the North Sea as biosurfactant-producing and n-alkane-degrading marine bacteria. These Gram-negative bacteria are peculiar as they cannot use carbohydrates and amino acids as growth substrates (Yakimov et al., 1998). When grown on n-alkanes, however, they produce biosurfactants which have been shown to be glucose lipids (Abraham et al., 1998). These bacteria become predominant in the microbial community in seawater containing crude oil. Figure 6 shows the result from a batch culture of natural seawater (without sterilization and without inoculation of exogenous microorganisms). At an early stage of cultivation, the *Alcanivorax* population exceeded more than 90% of the total population. Similar results were obtained when petroleum-degrading populations were examined in the marine-beach-simulating mesocosms described by Ishihara *et al.* (1995). Furthermore, outgrowth of an *Alcanivorax* population in a water column from the Sea of Japan was observed after the oil spill from the Russian tanker, *Nakhodka,* on January 1997 (Kishira, Kasai and Shutsubo, unpublished results).

Future Prospects

The bioremediation studies after the oil spill from the *Exxon Valdez* demonstrated the promise of this technology for cleaning up a petroleum-contaminated shoreline. It has been estimated that a beach damaged by oil spill could be returned to its natural condition in as little as two to five years with bioremediation treatment, otherwise it would take ten years or more to reach this condition. Ten years after the *Exxon Valdez* accident, oil residues still remain subsurface in the spill-affected bays where contamination problems persist and affect fishery activities. This fact clearly indicates the necessity for further development of bioremediation technologies to manage marine oil pollution. The microorganisms which play the major roles in the bioremediation of offshore sediment, and their nutrient requirements (carbon, nitrogen, phosphorous, etc.) and environmental requirements (oxygen or an alternative electron acceptor, temperature, redox potential, salinity, pH, etc.) should be determined. Such knowledge would allow the manipulation of environmental factors that may limit or prevent the biodegradation of petroleum in marine sediment. As mentioned earlier, bioaugmentation or seeding would be useful to enhance the cleanup process of oil-contaminated sediment.

Undertaking bioremediation cannot be done without the agreement of local communities. Their concerns about bioremediation, a relatively new technology, should be mitigated by scientists who can explain to the local people the results of hydrocarbon contamination tests and microorganism tests, especially in regard to risk assessment. The identification of the *Alcanivorax* group as a major population that arose during a bioremediation treatment would be useful information to help convince people that the propagation of harmful microorganisms would not occur during bioremediation.

Acknowledgements

The work in the authors' laboratory was supported by a grant from the New Energy and Industrial Technology Development Organization. We thank Ryu-ichiro Kurane for his encouragement.

References

Abraham, W.R., Meyer, H., and Yakimov, M. 1998. Novel glycine containing glucolipids from the alkanes using bacterium *Alcanivorax borkumensis*. Biochim. Biophys. Acta 1393: 57-62.

Atlas, R.M., and Bartha, R. 1993. Microbial Ecology: Fundamentals and Applications. 3rd ed. Addison Wesley Longman Publishers, Amsterdam.

Bragg, J.R., Prince, R.C., Harner, E.J., and Atlas, R.M. 1994. Effectiveness of bioremediation for the *Exxon Valdez* oil spill. Nature. 368: 413-418.

Breese, K., Boll, M., Alt-Morbe, J., Schagger, H., and Fuchs, G. 1998. Genes coding for the benzoyl-CoA pathway of anaerobic aromatic metabolism in the bacterium *Thauera aromatica*. Eur. J. Biochem. 256: 148-154.

Bregnard, T.P.-A., Haner, A., Hohener, P., and Zeyer, J. 1997. Anaerobic degradation of pristane in nitrate-reducing microcosms and enrichment cultures. Appl. Environ. Microbiol. 63: 2077-2081.

Coates, J.D., Woodward, J., Allen, J., Philp, P., and Lovley, D.R. 1997. Anaerobic degradation of polycyclic aromatic hydrocarbons and alkanes in petroleum-contaminated marine harbor sediments. Appl. Environ. Microbiol. 63: 3589-3593.

Coschigano, P.W., Wehrman, T.S., and Young, L.Y. 1998. Identification and analysis of genes involved in anaerobic toluene metabolism by strain T1: putative role of a glycine free radical. Appl. Environ. Microbiol. 64: 1650-1656.

Egland, P.G., Pelletier, D.A., Dispensa, M., Gibson, J., and Harwood, C.S. 1997. A cluster of bacterial genes for anaerobic benzene ring biodegradation. Proc. Natl. Acad. Sci. USA. 94: 6484-6489.

Ehrhardt, M., and Weber, R.R. 1991. Formation of low molecular weight carbonyl compounds by sensitized photochemical decomposition of aliphatic hydrocarbons in seawater. Fresenjus J. Anal. Chem. 339: 772-776.

Evans, W.C., and Fuchs, G. 1988. Anaerobic degradation of aromatic compounds. Annu. Rev. Microbiol. 42: 289-317.

Floodgate, G. 1984. The fate of petroleum in marine ecosystems. In: Petroleum Microbiology. R. M. Atlas, ed. Macmillan Publishing Co., New York. p. 355-398.

Foght, J.M., Gutnick, D.L., and Westlake, W. S. 1989. Effect of Emulsan on biodegradation of crude oil by pure and mixed bacteria cultures. Appl. Environ. Microbiol. 55: 36-42.

Gough, M.A., and Rowland, S.J. 1990. Characterization of unresolved complex mixtures of hydrocarbons in petroleum. Nature. 344: 648-650.

Harayama, S. 1997. Polycyclic aromatic hydrocarbon bioremediation design. Curr. Opin. Biotechnol. 8: 268-273.

Harayama, S., Kok, M., and Neidle, E.L. 1992. Functional and evolutionary relationships among diverse oxygenases. Annu. Rev. Microbiol. 46: 565-601.

Ishihara, M., Sugiura, K., Asaumi, M., Goto, M., Sasaki, E. and Harayama, S. 1995. Oil degradation in microcosms and mesocosms. In: Microbial Processes for Bioremediation. R. E. Hincheee, F. J. Brockman and C. M. Vogel, eds. Battelle Press, Columbus, Ohio. p. 101-116.

Johnson, G.R., and Olsen, R.H. 1997. Multiple pathways for toluene degradation in *Burkholderia* sp. strain JS150. Appl. Environ. Microbiol. 63: 4047-4052.

Karlsen, D.A., and Larter, S.R. 1991. Analysis of petroleum fractions by TLC-FID: applications to petroleum reservoir description. Org. Geochem. 17: 603-617.

Kennicutt, M.C. 1988. The effect of biodegradation on crude oil bulk and molecular composition. Oil & Chem. Pollut. 4: 89-112.

Killops, S.D., and Al-Juboori, M.A.H.A. 1990. Characterisation of the unresolved complex mixture (UCM) in the gas chromatograms of biodegraded petroleums. Org. Geochem. 15: 147-160.

Markovetz, A.J., and Kallio, R.E. 1971. Subterminal oxidation of aliphatic hydrocarbons by microorganisms. CRC Crit. Rev. Microbiol. 1: 225-237.

Maeng, J.H., Sakai, Y., Tani, Y., and Kato, N. 1996. Isolation and characterization of a novel oxygenase that catalyzes the first step of *n*-alkane oxidation in *Acinetobacter* sp. strain M-1. J. Bacteriol. 178: 3695-3700.

Matar, S., and Hatch, L.F. 1994. Chemistry of Petrochemical Processes. Gulf Publishing Co., Houston, Texas.

Morgan, P., and Watkinson, R.J. 1994. Biodegradation of components of petroleum. In: C. Ratledge, ed. Biochemistry of Microbial Degradation. Kluwer Academic Publishers, Dordrecht, Netherlands. pp. 1-31.

Morikawa, M., Kanemoto, M., and Imanaka, T. 1996. Biological oxidation of alkane to alkene under anaerobic conditions. J. Ferment. Bioeng. 82: 309-311.

Myhr, M.B., Schou, L., Skjetne, T., and Krane, J. 1989. Characterization of asphaltenes and co-precipitated material from a Californian crude oil. Org. Geochem. 16: 931-941.

Nakajima, K., Sato, A., Takahara, Y., and Iida, T. 1985. Microbial oxidation of isoprenoid alkanes, phytane, norpristane and farnesane. Agric. Biol. Chem. 49: 1993-2002.

Nicodem, D. E., Fernandes, M. C. Z., Guedes, C. L. B., and Correa, R. J. 1997. Photochemical processes and the environmental impact of petroleum spills. Biogeochem. 39: 121-138.

Oudot, J. 1984. Rates of microbial degradation of petroleum components as determined by computerized capillary gas chromatography and computerized mass spectrometry. Mar. Env. Res. 13: 277-302.

Prince, R.C. 1993. Petroleum spill bioremediation in marine environments. Crit. Rev. Microbiol. 19: 217-242.

Pritchard, P.H., and Costa, C.F. 1991. EPA's Alaska oil spill bioremediation project. Environ. Sci. Technol. 25: 372-379.

Readman, J.W., Fowler, S.W., Villeneuve, J.-P., Cattini, C., Oregioni, B., and Mee, L.D. 1992. Oil and combustion-product contamination of the Gulf marine environment following the war. Nature. 358: 662-665.

Rontani, J.F., Bosser-Joulak, F., Rambeloarisoa, E., Bertrand, J.E., Giusti, G., and Faure, R. 1985. Analytical study of Asthart crude oil asphaltenes biodegradation. Chemosphere. 14: 1413-1422.

Rosenberg, E., and Ron, E.Z. 1997. Bioemulsans: microbial polymeric emulsifiers. Curr. Opin. Biotechnol. 8: 313-316.

Saito, A., Iwabuchi, T., and Harayama, S. 1999. Characterization of genes for enzymes involved in the phenanthrene degradation in *Nocardioides* sp. KP7. Chemosphere. 38: 1331-1337.

Sasaki, T., Maki, H., Ishihara, M., and Harayama, S. 1998. Vanadium as an internal marker to evaluate microbial degradation of crude oil. Environ. Sci. Technol. 22: 3618-3621.

Scheller, U., Zimmer, T., Becher, D., Schauer, F., and Schunck, W. H. 1998. Oxygenation cascade in conversion of *n*-alkanes to α,ω-dioic acids catalyzed by cytochrome P450 52A3. J. Biol. Chem. 273: 32528-32534.

Shanklin, J., Achim, C., Schmidt, H., Fox, B.G., and Munck, E. 1997. Mossbauer studies of alkane omega-hydroxylase: evidence for a diiron cluster in an integral-membrane enzyme. Proc. Natl. Acad. Sci. USA. 94: 2981-2986.

Sugiura, K., Ishihara, M., and Harayama, S. 1997. Comparison of the biodegradability of four crude oil samples by a single bacterial species and by a microbial population. Environ. Sci. Technol. 31: 45-51.

Tjessen, K., Pedersen, D., and Aaberg, A. 1984. On the environmental fate of a dispersed Ekofisk crude oil in sea-immersed plastic columns. Water Res. 9: 1129-1136.

van Beilen, J.B., Wubbolts, M.G., and Witholt, B. 1994. Genetics of alkane oxidation by *Pseudomonas oleovorans*. Biodegradation. 5: 161-174.

Wang, Z., Fingas, M., Blenkinsopp, S., Sergy, G., Landriault, M., Sigouin, L., Foght, J., Semple, K., and Westlake, D. W. S. 1998. Comparison of oil composition changes due to biodegradation and physical weathering in different oils. J. Chromatography. A 809: 89-107.

Weiner, J.M., and Lovley, D.R. 1998. Anaerobic benzene degradation in petroleum-contaminated aquifer sediments after inoculation with a benzene-oxidizing enrichment. Appl. Environ. Microbiol. 64: 775-778.

Whyte, L.G., Hawari, J., Zhou, E., Bourbonniere, L., Inniss, W.E., and Greer, C.W. 1998. Biodegradation of variable-chain-length alkanes at low temperatures by a psychrotrophic *Rhodococcus* sp. Appl. Environ. Microbiol. 64: 2578-2584.

Yakimov, M.M., Golyshin, P.N., Lang, S., Moore, E.R., Abraham, W.R., Lunsdorf, H., and Timmis, K. N. 1998. *Alcanivorax borkumensis* gen. nov., sp. nov., a new, hydrocarbon-degrading and surfactant-producing marine bacterium. Int. J. Syst. Bacteriol. 48: 339-348.

Zhang, Y., and Miller, R.M. 1992. Enhanced octadecane dispersion and biodegradation by a *Pseudomonas* rhamnolipid surfactant (biosurfactant). Appl. Environ. Microbiol. 58: 3276-3282.

10

Marine *Bacillus* Spores as Catalysts for Oxidative Precipitation and Sorption of Metals

Chris A. Francis and Bradley M. Tebo

Marine Biology Research Division and
Center for Marine Biotechnology and Biomedicine,
Scripps Institution of Oceanography,
University of California, San Diego,
La Jolla, CA 92093-0202, USA

Abstract

The oxidation of soluble manganese(II) to insoluble Mn(III,IV) oxide precipitates plays an important role in the environment. These Mn oxides are known to oxidize numerous organic and inorganic compounds, scavenge a variety of other metals on their highly charged surfaces, and serve as electron acceptors for anaerobic respiration. Although the oxidation of Mn(II) in most environments is believed to be bacterially-mediated, the underlying mechanisms of catalysis are not well understood. In recent years, however, the application of molecular biological approaches has provided new insights into these mechanisms. Genes involved in Mn oxidation were first identified in our model organism, the marine *Bacillus* sp. strain SG-1, and subsequently have been identified in two other phylogenetically distinct organisms, *Leptothrix discophora* and *Pseudomonas putida*. In all three cases, enzymes related to multicopper oxidases appear to be involved, suggesting that copper may play a universal role in Mn(II) oxidation. In addition to catalyzing an environmentally important process, organisms capable of Mn(II) oxidation are potential candidates for the removal, detoxification, and recovery of metals from the environment. The Mn(II)-oxidizing spores of the marine *Bacillus* sp. strain SG-1 show particular promise, due to their inherent physically tough nature and unique capacity to bind and oxidatively precipitate metals without having to sustain growth.

Introduction

Microorganisms capable of manganese(II) oxidation have been recognized since the beginning of the 20th century (Jackson, 1901) but, even today, the underlying mechanisms and biological function of this process remain poorly understood. Despite a century of isolating and characterizing an amazing diversity of Mn(II)-oxidizing bacteria from a wide variety of environments, only recently has significant progress been made towards elucidating the mechanisms for enzymatic Mn(II) oxidation. This progress has been due primarily to the application of molecular and biochemical approaches to the study of bacterial Mn(II) oxidation.

The primary focus of this chapter is to review our current view of the mechanism for Mn(II) oxidation of the marine *Bacillus* sp. strain SG-1, with particular emphasis on the molecular genetic and biochemical aspects. In addition, comparisons with two other model bacterial Mn(II)-oxidation systems allow us to speculate regarding a more universal mechanism of Mn(II) oxidation. Finally, we review the unique metal binding and oxidation properties of SG-1 spores which make them attractive candidates for biotechnological applications, such as the bioremediation of metal pollution.

Background on Manganese(II) Oxidation

General Chemistry of Manganese

Manganese (Mn) is an essential nutrient for all living organisms, serving as a cofactor in a variety of enzymes (Larson and Pecoraro, 1992), including superoxide dismutase and the active site of photosystem II. Manganese is the second most abundant transition metal, behind iron, in the earth's crust and the fifth most abundant metal on the surface of the earth. Although Mn can occur in oxidation states ranging from 0 to +7, the +2, +3, and +4 oxidation states are most relevant under natural environmental conditions. In nature, Mn is generally found as reduced soluble or adsorbed Mn(II) and as highly insoluble Mn(III) and Mn(IV) oxides and oxyhydroxides, which appear as brownish-black precipitates. Mn(IV) minerals are ultimately the most thermodynamically stable form in nature.

Abiotic Mn Oxidation

The oxidation of soluble Mn(II) to Mn(III,IV) oxides is a thermodynamically favorable, but kinetically slow, reaction at neutral pH. Because of this, Mn(II) oxidation in natural systems, such as groundwater and surface waters, often proceeds at very slow rates in the absence of bacteria (Diem and Stumm, 1984; Nealson *et al.*, 1988). Abiotic chemical oxidation of Mn(II) generally only occurs under extreme conditions within a few weeks to months. In marine environments, soluble Mn can vary between 10^{-9} M in seawater to 10^{-4} M in pore waters of some sediments (Rosson and Nealson, 1982). Mn(II) oxidation is autocatalytic, with the Mn(oxyhydr)oxide products adsorbing Mn(II) and catalyzing its further oxidation. In addition, a variety of other surfaces like Fe oxides and silicates also catalyze Mn(II) oxidation. Mn oxides play an important role in the marine environment, where they are known to oxidize a number of organic and inorganic compounds, serve as electron acceptors for anaerobic bacteria, and scavenge many other metals (*e.g.*, Cu, Co, Cd, Ni, and Zn) on their highly charged surfaces.

Biological Mn Oxidation

Although the production of Mn oxides in most environments is considered to be predominantly microbially mediated (Nealson *et al.*, 1988), the mechanisms of catalysis (and biological function) are poorly understood. Mn(II)-oxidizing organisms are widely distributed in nature and occur wherever soluble Mn(II) species occur, from marine and freshwaters, to sediments, soils, and desert varnish (Ghiorse, 1984). Certain algae, yeast, and fungi have been shown to catalyze Mn(II) oxidation, but bacteria are believed to be the most important Mn(II)-oxidizing organisms in aquatic environments (Tebo *et al.*, 1997; Tebo, 1998). As a group, Mn(II)-oxidizing bacteria are phylogenetically diverse. Based on 16S rRNA sequencing, all Mn(II)-oxidizers analyzed to date have fallen within the Gram-positive or *Proteobacteria* branches of the Domain Bacteria (Tebo *et al.*, 1997). In addition, all of the Gram-negative organisms have fallen within the α, β, and γ *Proteobacteria*.

There are two general mechanisms of Mn oxidation (Nealson *et al.*, 1989) which can be operationally described as indirect or direct. Indirect oxidation may occur via an increase in pH or E_h, while direct oxidation generally occurs via the active binding and oxidation of Mn(II) by an enzyme. Mn(II)-oxidizing activity has been reported in cell-free extracts of many bacteria (Ehrlich, 1968; Douka, 1977; Jung and Schweissfurth, 1979; Douka, 1980), but the specific Mn-oxidizing components have only been characterized in a few cases, namely: *Leptothrix discophora*, *Pseudomonas putida*, and our model organism, the marine *Bacillus* sp. strain SG-1.

Leptothrix sp. are sheath-forming organisms which are ubiquitous in wetlands, iron seeps, and springs around the world. *L. discophora* is characterized by the precipitation of both iron and manganese oxides on its sheaths. The sheathless mutant strain SS-1 excretes a manganese-oxidizing factor, normally associated with the sheath, into the culture medium. This ~110 kDa protein, designated MofA, is capable of forming a Mn oxide band in SDS-PAGE gels incubated in $MnCl_2$. MofA was the first Mn-oxidizing protein to be purified and partially characterized (Adams and Ghiorse, 1987; Boogerd and deVrind, 1987). The oxidizing activity is inhibited by cyanide, azide, *o*-phen-anthroline, mercuric chloride, and pronase. The gene *mofA*, encoding the putative Mn(II)-oxidizing protein of SS-1, was recently cloned and sequenced (Corstjens *et al.*, 1997), which revealed that the encoded protein sequence shares significant similarity with multicopper oxidases (see below). However, further analysis of the molecular mechanism of Mn oxidation of *L. discophora* has been hampered by the current lack of genetic tools for use in these organisms.

Pseudomonas putida is a ubiquitous freshwater and soil bacterium and, thus, provides an excellent model system for studying bacterial Mn oxidation. The closely related strains MnB1 and GB-1 have been intensively studied in recent years. Upon reaching stationary phase, these organisms oxidize Mn(II) to Mn(IV) oxyhydroxides which are precipitated on the cell surface. Previous studies demonstrated that MnB1 produces a soluble Mn(II)-oxidizing protein late in logarithmic phase (Jung and Schweissfurth, 1979; DePalma, 1993). More recent biochemical studies with GB-1 resulted in the partial purification and characterization of two Mn(II)-oxidizing factors with estimated molecular weights of 180 kDa and 250 kDa (Okazaki *et al.*, 1997). The Mn-oxidizing activity of these factors is sensitive to azide and mercuric chloride, and inhibited by cyanide, EDTA, Tris, and *o*-phenanthroline. Unlike MofA of *L. discophora*, the Mn(II)-oxidizing factors of GB-1 are more sensitive to SDS and only produce Mn oxide bands in native polyacrylamide gels (lacking SDS). Rather than the existence of two distinct Mn-oxidizing proteins, it is more likely that the Mn-oxidizing protein(s) isolated are part of a larger complex which degrades into smaller fragments that retain activity (Okazaki *et al.*, 1997).

In contrast to *L. discophora*, there are a variety of well-developed genetic tools available for molecular genetic analysis of *Pseudomonas* species. Recent studies have used transposon mutagenesis to identify genes involved in Mn oxidation in both *P. putida* strain MnB1 and GB-1 (Caspi *et al.*, 1998; de Vrind *et al.*, 1998). In both studies, genes involved in the biogenesis and maturation of *c*-type cytochromes were found to be essential for Mn oxidation. However, cytochromes alone are not thought to be sufficient for catalyzing the oxidation of manganese. In GB-1, a gene encoding a multicopper oxidase, designated *cumA*, was found to be essential for Mn-oxidation (Brouwers *et al.*, 1999). In addition, small amounts of Cu^{2+} were found to increase the Mn(II)-oxidizing activity of wild-type cells by a factor of 5. Thus, it has been proposed that this Cu-dependent oxidase is an important constituent of the oxidizing complex and may directly oxidize Mn(II). The importance of copper in the mechanism of bacterial manganese oxidation will be further discussed later in this chapter.

Marine *Bacillus* sp. Strain SG-1

General Properties
Spore-forming *Bacillus* species can be a significant component of the total colony-forming bacteria in certain aquatic environments (20 to 40%) and sediments (up to 80%) (Bonde, 1981). Within the genus *Bacillus*, a variety of organisms are known to oxidize Mn(II). Some oxidize Mn(II) during vegetative growth (Ehrlich and Zapkin, 1985; Ehrlich, 1996) or only during the onset of sporulation (Vojak *et al.*, 1984), but there is a major group, at least in marine environments, that produce mature spores that oxidize Mn(II) (Lee, 1994). In fact, a considerable portion (17-33%) of the spore-forming bacteria isolated from coastal surface sediments of Mission Bay and San Diego Bay, California, were found to produce Mn(II)-oxidizing spores (Lee and Tebo, unpublished).

The marine *Bacillus* sp. strain SG-1 was isolated from a Mn-coated sand grain that was obtained from a shallow marine sediment off Scripps pier (Nealson and Ford, 1980). This organism produces metabolically dormant spores that bind and oxidize Mn(II), thereby becoming encrusted with Mn oxide (Figure 1). SG-1 spores are also capable of binding a variety of other heavy metals such as copper, cadmium, zinc, nickel, and cobalt (the latter of which is also oxidized) (Tebo and Lee, 1993; Lee and Tebo,

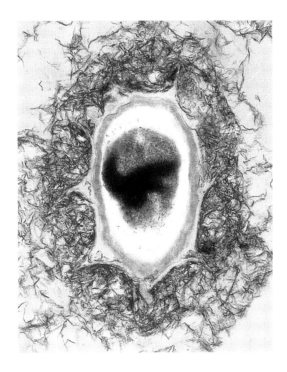

Figure 1. Spores of *Bacillus* sp. Strain SG-1
Transmission electron micrograph of a thin section of the metal-oxidizing spores of the marine *Bacillus* sp. strain SG-1. The spores are coated with manganese oxides. Approximate spore size: 1.25 x 0.75 µm.

1994; Tebo, 1995). The vegetative cells of SG-1, on the other hand, do not oxidize Mn and have actually been shown to reduce Mn oxide under oxygen limiting conditions (de Vrind *et al.*, 1986a). This suggests that one possible purpose of Mn oxidation by these spores is to store up Mn oxides as an electron acceptor for growth under low oxygen or anaerobic conditions, upon germination in the sediments (Tebo, 1983; de Vrind *et al.*, 1986a).

Biochemistry

Manganese oxidation by SG-1 spores occurs over a wide range of environmental conditions including: metal concentration (<nM to >mM), temperature (<3 °C to > 70 °C), pH (>6.5), and osmotic strength (from distilled water to seawater) (Rosson and Nealson, 1982). In fact, the spores can even be rendered non-germinable with glutaraldehyde, formaldehyde, or UV light, and still retain Mn oxidizing activity (Rosson and Nealson, 1982). The oxidizing activity of the spores is heat labile and is poisoned by the metalloprotein inhibitors azide, cyanide, and mercuric chloride (Rosson and Nealson, 1982). Transmission electron microscopy demonstrated that the Mn oxide is precipitated on the ridged outermost spore layer (Tebo, 1983). Spore coat preparations, processed to retain all the outer layers and remove the spore contents, were shown to retain full oxidizing activity (de Vrind *et al.*, 1986b). These results suggested that a protein component of the outermost spore layer, either the spore coat or exosporium, is responsible for catalyzing the oxidation of manganese.

The spore coat is a highly cross-linked structure which gives the spore resistance to chemical attack and mechanical disruption (Warth, 1978; Driks,

1999). An additional layer found in some, but not all, spores is termed the exosporium. The exosporium is a loose-fitting outermost layer composed of protein, lipid, and carbohydrate, and has no known function (Matz *et al.*, 1970; Tipper and Gauthier, 1972). Although this layer has been hypothesized to play a protective role, this is somewhat controversial since it is not found in spores of all species. Unlike the spore coat, there is very little information available regarding the exosporium at the genetic, biochemical, or developmental level. Recent studies in our laboratory suggest that the Mn(II)-oxidizing activity of SG-1 spores is localized to an exosporium (Francis *et al.*, 1997).

Over the years, attempts have been made to isolate the Mn(II)-oxidizing protein(s) by extracting proteins from SG-1 spores, separating them by SDS-PAGE, and incubating the gels with Mn(II) (Tebo *et al.*, 1988). A high molecular weight Mn-oxidizing band (~205 kDa) has occasionally been observed in gels. Re-extraction of this band, followed by SDS-PAGE, and Coomassie staining revealed that it was composed of several proteins. However, these experiments are difficult to reproduce, from experiment to experiment, possibly due to damaging of the Mn-oxidizing factors during extraction, or because several components that are separated during electrophoresis may be required for activity.

Genetics
Due to the difficulties in consistently recovering Mn(II)-oxidizing activity from spores for biochemical studies, our laboratory employed a molecular genetic approach to study Mn oxidation by SG-1. Methods for plasmid transformation and transposon mutagenesis were developed for SG-1 (van Waasbergen *et al.*, 1993). Using the temperature sensitive plasmid pLTV1, which carries Tn*917*, a promoterless *lacZ* gene, and an *Escherichia coli* replicon, 27 independent non-oxidizing, but still sporulating, mutants were isolated. Out of the 27 mutants, 18 of the insertions turned out to map within a contiguous cluster of seven genes, the *mnx* genes (van Waasbergen *et al*, 1996). This work was the first report to identify genes involved in Mn oxidation, as well as the first to describe a genetic system developed for a marine Gram-positive bacterium.

The *mnx* gene cluster appears to be organized in an operon (Figure 2) which is preceded by a potential recognition site for the sporulation, mother-cell-specific, RNA polymerase sigma factor, σ^K. Consistent with this, measurement of ß-galactosidase activity from a Tn*917-lacZ* insertion in *mnxD* showed expression at mid- to late sporulation (approximately stage IV to V of sporulation). Spores of nonoxidizing mutants appeared unaffected with respect to their temperature and chemical resistance properties as well as germination characteristics. However, in some of the mutants, transmission electron microscopy revealed slight alterations in the ridged outermost spore layer, consistent with the localization of Mn(II)-oxidizing activity to this layer.

Possible Mechanism of Mn Oxidation
Sequence analysis of the *mnx* gene cluster revealed that three of the encoded proteins (MnxA, MnxB, and MnxE) were predicted to be highly hydrophobic, while only two of the proteins (MnxC and MnxG) showed significant similarity

to other proteins in the databases. MnxG is a predicted 138 kDa protein which shows similarity to the family of multicopper oxidases (Figure 3), a diverse group of proteins that utilize multiple copper ions as cofactors in the oxidation of a variety of substrates (Ryden and Hunt, 1993). Members of this family include ascorbate oxidase (from squash and cucumber), laccase (from plants and fungi), ceruloplasmin (from vertebrates), FET3 (from yeast), and CopA (a copper resistance protein from *Pseudomonas syringae*). Of these proteins, only ceruloplasmin and FET3 are known to oxidize a metal, Fe(II), as a substrate.

Multicopper oxidases are a unique class of enzymes which can be defined by their spectroscopy, sequence homology, and reactivity (Solomon, 1996). All multicopper oxidases contain copper ions of three spectroscopically distinct types ('blue' copper (or Type 1), Type 2, and Type 3) with the minimum functional unit containing at least one Type 1 site and a Type 2/Type 3 trinuclear cluster. The amino acids (histidine, cysteine, and methionine/leucine/phenylalanine) which make up each copper center come into close proximity to one another and coordinate copper. The Type 1 center accepts the initial electron from the substrate and shuttles it to the Type 2/Type 3 center which binds and reduces molecular oxygen:

$$\text{Substrate} \xrightarrow{e^-} \text{Type 1} \xrightarrow{e^-} \{\text{Type 2} + (\text{Type 3})_2\} \xrightarrow{e^-} O_2$$

Only multicopper oxidases and cytochrome oxidases are known to couple the four electron reduction of O_2 to H_2O with the oxidation of substrate. In the well-characterized multicopper oxidases, the substrate is oxidized by one electron. Thus, if Mn(II) oxidation is, indeed, catalyzed by a multicopper oxidase, it is most likely that Mn(II) is oxidized by sequential one electron transfers in which Mn(III) is a transient intermediate.

Both X-ray crystallography and comparative sequence analysis have demonstrated that multicopper oxidases possess a distinctive subdomain structure (Solomon *et al.*, 1996). Laccase, ascorbate oxidase, and FET3, all appear to have three domains while, the larger enzyme, ceruloplasmin has six domains. These copper enzymes all exhibit significant internal homology among the subdomains, suggesting that they all arose from a common ancestor by gene duplication (Ryden and Hunt, 1993; Solomon *et al.*, 1996).

MnxG shares significant similarity with the multicopper oxidases, particularly in regions surrounding the conserved copper binding regions. Based on size and subdomain structure, MnxG appears to be most similar to the Fe(II)-oxidizing protein, ceruloplasmin, containing six subdomains.

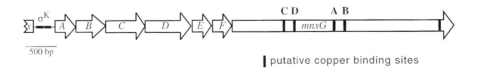

Figure 2. The *mnx* Gene Cluster
The organization of the *mnx* gene cluster of the marine *Bacillus* sp. strain SG-1 based on DNA sequence analysis. *mnxG* encodes the putative Mn(II)-oxidizing protein which shares significant similarity with multicopper oxidases, particularly in the regions of copper binding (boxed areas). The amino acid sequences of the copper binding sites designated with the letters A-D are shown in Figure 3. σ^K represents the putative -35 and -10 consensus promoter sequences which precede this operon.

A

MnxG	527	M	**H**	I	**H**	F	V
MofA	304	I	**H**	L	**H**	G	G
CumA	95	I	**H**	W	**H**	G	I
Asox	94	I	**H**	W	**H**	G	I
Lacc	78	V	**H**	W	**H**	G	L
Hcer	119	F	**H**	S	**H**	G	I
Fet3	8	M	**H**	F	**H**	G	L
CopA	99	I	**H**	W	**H**	G	I
			2		3		

B

MnxG	572	F	F	**H**	D	**H**
MofA	384	W	Y	**H**	D	**H**
CumA	136	W	Y	**H**	P	**H**
Asox	137	F	Y	**H**	G	**H**
Lacc	121	W	Y	**H**	S	**H**
Hcer	178	I	Y	**H**	S	**H**
Fet3	52	W	Y	**H**	S	**H**
CopA	140	W	Y	**H**	S	**H**
				3		3

C

MnxG	281	**H**	V	F	**H**	Y	**H**	V	**H**
MofA	1174	**H**	P	V	**H**	F	**H**	L	L
CumA	391	**H**	P	I	**H**	L	**H**	G	M
Asox	480	**H**	P	W	**H**	L	**H**	G	**H**
Lacc	508	**H**	P	I	**H**	K	**H**	G	N
Hcer	994	**H**	T	V	**H**	F	**H**	G	**H**
Fet3	341	**H**	P	F	**H**	L	**H**	G	**H**
CopA	542	**H**	P	I	**H**	L	**H**	G	M
		1			2		3		

D

MnxG	334	**H**	**C**	**H**	L	Y	P	**H**	F	G	I	**G** **M**
MofA	1279	**H**	**C**	**H**	I	L	G	**H**	E	E	N	D F
CumA	442	**H**	**C**	**H**	V	I	D	**H**	M	E	T	**G** L
Asox	542	**H**	**C**	**H**	I	E	P	**H**	L	H	M	**G** **M**
Lacc	585	**H**	**C**	**H**	I	A	S	**H**	Q	M	G	**G** **M**
Hcer	1039	**H**	**C**	**H**	V	T	D	**H**	I	H	A	**G** **M**
Fet3	411	**H**	**C**	**H**	I	E	W	**H**	L	L	Q	**G** L
CopA	590	**H**	**C**	**H**	L	L	Y	**H**	M	E	M	**G** **M**
		3	1	3				1				1

Figure 3. Copper-binding Sites in Multicopper Oxidases
Amino acid alignment of the copper-binding sites in MnxG, MofA, CumA, and other multicopper oxidases. The letters A-D correspond to the copper binding sites shown in Figure 2. Abbreviations: Asox, ascorbate oxidase (cucumber and squash); Lacc, laccase (fungi); Hcer, human ceruloplasmin; FET3, an iron oxidizing/ transport protein in yeast; and CopA, a copper-resistance protein from *Pseudomonas syringae*. The amino acids conserved among the different proteins are shaded and the copper-binding residues are numbered according to the spectroscopic type of copper they potentially help coordinate.

Azide, a potent inhibitor of multicopper oxidases that acts by bridging the Type 2 and Type 3 copper atoms, has also been found to inhibit Mn(II) oxidation by SG-1 spores. Conversely, small amounts of copper actually enhance the rate of Mn(II) oxidation by the spores (van Waasbergen *et al.*, 1996). The sequence similarity of MnxG to multicopper oxidases, combined with the copper-enhancement and azide-inhibition of Mn(II) oxidation, suggests that MnxG may function like a copper oxidase and directly oxidize manganese.

Although MnxG may be the only Mnx protein directly involved in Mn oxidation, it is possible that one or more of the other Mnx proteins may also be required for activity. In particular, MnxC shares significant similarity with several proteins involved in multicomponent oxidoreductase systems, suggesting that MnxC and MnxG might also part of such a system. MnxC is a predicted 22 kDa protein that has a putative N-terminal signal sequence, indicating that it may be associated with a membrane. It shares sequence similarity with a number of cell surface and multicomponent oxidoreductase-associated proteins which all share a C-XXX-C motif. One of these proteins, an 18 kDa protein in the mercury resistance operon of *Staphylococcus aureus* (Laddaga *et al.*, 1987), has a thioredoxin motif surrounding these cysteine residues (C-XX-C), suggesting that these residues may exhibit redox activity and be involved in the formation of disulfide bonds (Ellis *et al.*, 1992). An alternative, and perhaps more intriguing, function for these cysteine residues comes from the similarity of MnxC to two other proteins, SCO1 and SCO2 of *Saccharomyces cerevisiae*. These proteins were previously shown to play an essential role in the assembly of the mitochondrial cytochrome oxidase complex (Schulze and Roedel, 1989). More recently, the two cysteines of SCO1 have been suggested to bind and deliver copper to the copper-containing protein, cytochrome oxidase, thus conferring activity (Glerum *et al.*, 1996). By analogy, MnxC may be involved in delivering copper to the multicopper oxidase, MnxG, giving it activity. Interestingly, the multicopper oxidases, ceruloplasmin and FET3, both require additional proteins to deliver copper to them and, thus, confer oxidase activity (Stearman *et al.*, 1996). A possible association between MnxC and MnxG is supported by the recent localization of both of these proteins to the exosporium of SG-1 spores (Francis and Tebo, unpublished).

Role of Copper in Bacterial Mn Oxidation
The bacterial Mn oxidation systems that have been characterized at the molecular level in recent years all seem to be linked by the apparent use of copper as an essential enzymatic cofactor. Three otherwise unrelated organisms, a *Leptothrix*, a *Pseudomonas*, and a *Bacillus* species, appear to be utilizing enzymes related to multicopper oxidases for the oxidation of manganese. Despite their involvement in catalyzing the same reaction, these extracellular proteins have unique locations within their respective organisms: within an extracellular sheath, an outer membrane, and an outermost spore layer. None of these proteins share strong overall sequence similarity with one another, but they all contain the conserved copper-binding regions always found in multicopper oxidases. In addition to sequence homology, there is also biochemical evidence to support the role of copper in bacterial Mn(II) oxidation. First, the Mn(II)-oxidizing activity of all three of these systems is inhibited by azide, a potent inhibitor of multicopper oxidases. Second, copper has been shown to significantly enhance the rate of Mn(II)-oxidation in both the *Bacillus* sp. strain SG-1 and *P. putida* GB-1 (Brouwers *et al.*, 1999), but has not yet been thoroughly tested in *L. discophora* SS-1. Finally, it has recently been demonstrated that Mn(II) oxidation in yet another phylogenetically distinct organism, the prosthecate bacterium *Pedomicrobium* sp. ACM 3067, also appears to be catalyzed by a copper-dependent enzyme (Larsen *et al.*, 1999). Although the well-known multicopper oxidases have been shown to oxidize a wide variety of substrates, until recently, Fe(II) was the only known metal substrate. Thus, it is possible that bacterial Mn oxidases may constitute a new functional group of multicopper oxidases. However, definitive proof of this hypothesis awaits further biochemical and spectroscopic analysis of these Mn(II)-oxidizing enzymes.

Potential Biotechnological Applications

In addition to providing an excellent model system for studying the molecular and biochemical mechanisms of metal precipitation, SG-1 spores have a number of unique properties that make them attractive candidates for biotechnological applications, such as environmental remediation of metal pollutants (Figure 4).

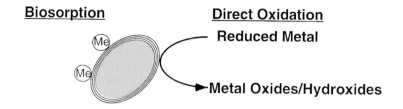

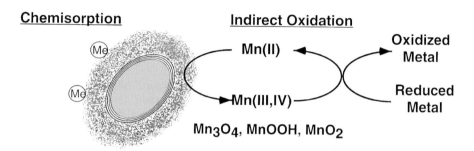

Figure 4. Schematic Representation of the Mechanisms by which SG-1 Spores can Either Adsorb or Oxidize Various Metals

Top: The spores can passively adsorb certain metals (where **Me** = Cu, Cd, Zn, Ni) on the charged spore surface (biosorption). The enzymatic activity of the outermost spore layer can also directly catalyze the oxidation of divalent metals such as Mn(II) and Co(II) (direct oxidation). **Bottom:** The highly charged Mn oxides which form on the spore surface are capable of nonspecifically adsorbing (chemisorption) a variety of metals (where **Me** = Cu, Co, Cd, Zn, Pb, radionuclides, etc.). The Mn oxides are also strong oxidants, capable of indirectly oxidizing many metals and organics (indirect oxidation).

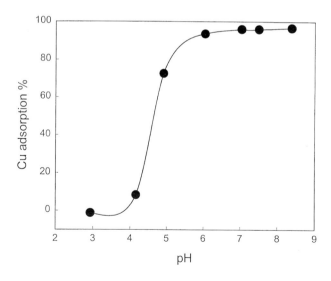

Figure 5. Cu(II) Adsorption by SG-1 Spores

Cu(II) adsorption by SG-1 spores as a function of pH [10^8 spores ml^{-1}, 0.01 M NaNO$_3$, 2 µM Cu(II)]. The adsorption value for 100% adsorption is 1.5 µmol m^{-2}. Reproduced with permission from He and Tebo (1998).

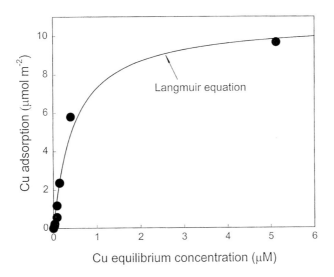

Figure 6. Cu(II) Adsorption Isotherm
Cu(II) adsorption isotherm obtained with SG-1 spores (10^8 spores ml^{-1}, 0.01 M NaNO$_3$, pH 7.0). The curve was obtained by fitting the data to the Langmuir equation. The results demonstrate that the spores have both a high affinity (K$_s$ = 0.48 µM) and a high adsorption capacity for Cu (10.77 µmol·m^{-2} of spore surface or 0.83 mmol·g^{-1} dried spores). Reproduced with permission from He and Tebo (1998).

Recent characterization of the surface chemistry and Cu(II) adsorption properties of SG-1 spores revealed that, in addition to actively binding and oxidizing Mn(II), they also have an extensive capacity for passively binding other metals (He and Tebo, 1998). The specific surface area of the spores was found to be around 74.7 m^{2}g^{-1}, a fairly high value in the range similar to metal (hydr)oxides and other clay minerals. Like most bacterial surfaces, the SG-1 spore surface has a net negative charge with a point of zero charge at pH 4.5. The surface was shown to be dominated by negatively charged sites, which are most likely carboxylate but also phosphate groups, consistent with the presence of both protein and carbohydrate in the outermost layer of the spores. Copper adsorption by SG-1 spores is rapid and complete within minutes, with adsorption starting at pH 3 and increasing with pH (Figure 5). The high surface area and surface site density of SG-1 spores is comparable to that of Fe, Mn, and Al mineral colloids, accounting for the fact that these spores have an extensive capacity for binding copper and other toxic metals on their surface (Figure 6). In fact, SG-1 spores have also been shown to bind both Cd(II) and Zn(II) (Tebo, 1995). The Cu(II) adsorption affinity coefficient (K) and the adsorption capacity (Γ_m) of the spores calculated from the Langmuir equation are 2.08 x 10^6 L·mol^{-1} and 10.77 µmol·m^{-2} respectively. The K value is simply the inverse of the substrate binding constant (K$_s$ = 0.48 µM) with which most biologists are familiar. The spore affinity for Cu(II) is 2-4 orders of magnitude greater (*i.e.* the K$_s$ is 2-4 orders of magnitude lower) than the affinities of Cu(II) determined for a variety of other biomasses, including fungi, bacteria and algae, or for an alginate gel (He and Tebo, 1998). The adsorption capacity is on the high end of the range observed for other types of biomass. Thus, in the absence of Mn, SG-1 spores may act as good passive adsorbents for the removal of metals and radionuclides from contaminated waters.

SG-1 spores also have the unique capacity to bind and oxidize cobalt (Tebo and Lee, 1993; Lee and Tebo, 1994), even in the absence of Mn(II) or preformed Mn oxides. Like Mn, the concentrations of Co in the environment rarely reach toxic levels. However, the radionuclide ^{60}Co is an activation product in radioactive wastes and has been identified as a priority pollutant at various Department of Energy sites in the United States. Since the oxidation of Co(II) results in the formation of solid Co(III)(oxy)hydroxide precipitates, the Co binding and oxidizing properties of SG-1 spores may be useful for dealing with ^{60}Co problems. The Co(II)-oxidizing properties of SG-1 spores are similar to those for Mn(II) oxidation, with oxidation occurring over a wide range of pH, temperature, and Co(II) concentrations (Lee and Tebo, 1994). Optimal Co(II) oxidation occurs around pH 8 and at 55 ° to 60 °C. Co(II) can be oxidized at the trace levels found in seawater all the way up to 100 mM, with the oxidation following Michaelis-Menton kinetics. Based on the kinetic studies, it appears that SG-1 spores have two oxidation systems for Co(II), a high-affinity-low rate system ($K_M = 3.3 \times 10^{-8}$ M; $V_{max} = 1.7 \times 10^{-15}$ $M^{-1}\cdot h^{-1}$) and a low-affinity-high-rate system ($K_M = 5.2 \times 10^{-6}$ M; $V_{max} = 8.9 \times 10^{-15}$ $M^{-1}\cdot h^{-1}$)(Lee and Tebo, 1994). It is likely that both Mn(II) and Co(II) oxidation occur at the same active site, since spores of transposon mutants within the *mnx* gene cluster do not oxidize Mn(II) or Co(II). The K_M for the high-affinity system (33 nM) suggests that SG-1 spores can remove metals (at least Co and Mn) to much lower levels than achieved by other chemical and biological procedures.

Mn oxides have long been recognized for their role in scavenging metals and radionuclides in the environment (Murray, 1975; Hem, 1978). The highly charged surfaces scavenge a variety of trace elements (*e.g.* Cu, Co, Cd, Zn, Pb) and radionuclides (*e.g.* ^{210}Pb, ^{60}Co), as well as Ra and Th isotopes, and can lead to the reduction in the concentration of soluble trace metals by several orders of magnitude. Thus, bacteria capable of catalyzing the precipitation of Mn oxides may be useful for application in the removal and recovery of toxic metals from the environment. SG-1 spores are particularly well suited for this purpose for a variety of reasons (Tebo *et al.*, 1998). The oxidation of Mn(II) can occur over a wide range of environmental conditions and can accumulate on the surface of spores up to approximately 6 times their own weight under ideal conditions. In addition, the rates of Mn oxidation by SG-1 spores at neutral pH are over 4-5 orders of magnitude faster than abiotic Mn oxidation rates (Hastings and Emerson, 1986). Biological Mn(II) oxidation has actually been employed as an alternative to chemical oxidation in the removal of excess Mn from drinking water (Mouchet, 1992) as well as for the removal of toxic contaminants from mine drainage (Mathur *et al.*, 1988). A recent study also demonstrated that biogenic Mn oxides produced by *Leptothrix discophora* SS-1 have significantly greater surface area and Pb adsorption capacity than abiotically produced Mn oxide (Nelson *et al.*, 1999). This extremely high trace metal adsorption capacity of biologically produced Mn oxides provides yet another advantage to employing Mn(II)-oxidizing bacteria for metal removal processes. Finally, the toxic metals adsorbed on the Mn oxides could be released by dissolving the oxides (*e.g.* with reducing agents) and the metals could be recovered and the spores recycled.

Conclusions and Future Directions

Bacteria play a central role in the biogeochemical cycling of metals in the environment, yet the molecular and biochemical mechanisms for most of these processes are not well understood. Clearly, the application of molecular biological approaches to the study of bacterial Mn(II) oxidation has transformed our understanding of this environmentally important process. However, it is important to recognize that this field is merely in its infancy and that major discoveries will surely be made in the very near future.

There are a number of important research avenues that should be pursued in future studies. Molecular analysis of other phylogenetically diverse Mn(II)-oxidizing bacteria using PCR primers and gene probes specific for known Mn(II) oxidation genes could reveal if multicopper oxidases are involved in all enzymatic Mn(II) oxidation systems. Functional gene probes for Mn(II) oxidation could then also be used to assess the distribution, abundance, and activities of Mn(II)-oxidizing organisms in the natural environment, even without the cultivation of organisms. Finally, more detailed biochemical and, especially, spectroscopic studies will be necessary to definitively elucidate the role of copper in the model Mn(II) oxidation systems. Overall, such studies should help further our understanding of the molecular basis (and function) of Mn oxidation, the factors which influence this environmentally important process, and how these organisms might be utilized to benefit society.

Acknowledgements

We are grateful for funding from the National Science Foundation (MCB94-07776 and MCB98-08915), the Collaborative UC/Los Alamos Research Program, and the University of California Toxic Substances Research and Training Program. C.A.F. was supported by a STAR Graduate Fellowship from the U.S. Environmental Protection Agency. We also acknowledge many undergraduate and graduate students, post-docs and technicians, including K. Casciotti, D. Edwards, L. He, Y. Lee, K. Mandernack, L. Park, and L. van Waasbergen, for their significant contributons to the studies of SG-1.

References

Adams, L.F., and Ghiorse, W.C. 1987. Characterization of extracellular Mn-oxidizing activity and isolation of a Mn-oxidizing protein from *Leptothrix discophora* SS-1. J. Bacteriol. 169: 1279-1285.
Bonde, G.J. 1981. *Bacillus* from marine habitats: allocation to phena established by numerical techniques. In: The Aerobic Endospore-forming Bacteria: Classification and Identification. R.C.W. Berkeley and M.

Goodfellow, eds. Academic Press, New York. p. 181-215.

Boogerd, R.C., and de Vrind, J.P.M. 1987. Manganese oxidation by *Leptothrix discophora* SS-1. J. Bacteriol. 169: 489-494.

Brouwers, G.J., de Vrind, J.P.M., Corstjens, P.L.A.M., Cornelis, P., Baysse, C., and de Vrind-de Jong, E.W. 1999. *CumA*, a gene encoding a multicopper oxidase, is involved in Mn^{2+}-oxidation in *Pseudomonas putida* GB-1. Appl. Environ. Microbiol. 65: 1762-1768.

Caspi, R., Haygood, M.G., and Tebo, B.M. 1998. *c*-type cytochromes and manganese oxidation in *Pseudomonas putida* MnB1. Appl. Environ. Microbiol. 64: 3549-3555.

Corstjens, P.L.A.M., de Vrind, J.P.M, Goosen, T., and de Vrind-de Jong, E.W. 1997. Identification and molecular analysis of the *Leptothrix discophora* SS-1 *mofA* gene, a gene putatively encoding a manganese-oxidizing protein with copper domains. Geomicrobiol. J. 14: 91-108.

DePalma, S.R. 1993. Manganese oxidation by *Pseudomonas putida*. Ph.D. thesis. Harvard University, Cambridge, Massachussetts.

de Vrind, J.P.M., Boogerd, F.C., and de Vrind-de Jong, E.W. 1986a. Manganese reduction by a marine *Bacillus* species. J. Bacteriol. 167: 30-34.

de Vrind, J.P.M, Brouwers, G.J., Corstjens, P.L.A., den Dulk, J., and de Vrind-de Jong, E.W. 1998. The cytochrome *c* maturation operon is involved in manganese oxidation in *Pseudomonas putida* GB-1. Appl. Environ. Microbiol. 64: 3556-3562.

de Vrind, J.P.M., de Vrind-de Jong, E.W., de Voogt, J.-W.H., Westbroek, P., Boogerd, F.C., and Rosson, R.A. 1986b. Manganese oxidation by spores and spore coats of a marine *Bacillus* species. Appl. Environ. Microbiol. 52: 1096-1100.

Diem, D., and W. Stumm. 1984. Is dissolved Mn^{2+} being oxidized by O_2 in the absence of Mn-bacteria or surface catalysts? Geochim. Cosmochim. Acta. 48: 1571-1573.

Douka, C. 1977. Study of the bacteria from manganese concretions. Soil. Biol. Biochem. 9: 89-97.

Douka, C. 1980. Kinetics of manganese oxidation by cell-free extracts of bacteria isolated from manganese concretions from soil. Appl. Environ. Microbiol. 39: 74-80.

Driks, A. 1999. *Bacillus subtilis* spore coat. Microbiol. Mol. Biol. Rev. 63: 1-20.

Ehrlich, H.L. 1968. Bacteriology of manganese nodules II. Manganese oxidation by cell-free extract from a manganese nodule bacterium. Appl. Microbiol. 16: 197-202.

Ehrlich, H. L. 1996. Geomicrobiology. Marcel Dekker, Inc., New York.

Ehrlich, H.L., and Zapkin, M.A. 1985. Manganese-rich layers in calcareous deposits along the western shore of the Dead Sea may have a bacterial origin. Geochem. J. 4: 207-221.

Ellis, L.B.M, Saurugger, J.C., and Woodward, C. 1992. Identification of the three-dimensional thioredoxin motif: related structure in the ORF3 protein in the *Staphylococcus aureus mer* operon. Biochem. 31: 4882-4891.

Francis, C.A., Casciotti, K.L., and Tebo, B.M. 1997. Manganese oxidizing activity localized to the exosporium of spores of the marine *Bacillus* sp.

strain SG-1. In: Abstracts of the 97th Annual Meeting of the American Society for Microbiology Miami, FL, May 1997. American Society for Microbiology, Washington, D.C. p. 499, Q-264.

Ghiorse, W.C. 1984. Biology of iron- and manganese-depositing bacteria. Annu. Rev. Microbiol. 38: 515-550.

Glerum, D.M., Shtanko, A., and Tzagoloff, A. 1996. SCO1 and SCO2 act as high copy suppressors of a mitochondrial copper recruitment defect in *Saccharomyces cerevisiae*. J. Biol. Chem. 271: 20531-20535.

Hastings, D., and Emerson, S. 1986. Oxidation of manganese by spores of a marine bacillus: kinetic and thermodynamic considerations. Geochim. Cosmochim. Acta. 50: 1819-1824.

He, L.M., and Tebo, B.M. 1998. Surface charge properties and Cu(II) adsorption by spores of the marine *Bacillus* sp. strain SG-1. Appl. Environ. Microbiol. 64: 1123-1129.

Hem, J.D. 1978. Redox processes at surfaces of manganese oxide and their effects on aqueous metal ions. Chem. Geol. 21: 199-218.

Jackson, D.D. 1901. The precipitation of iron, manganese, and aluminum by bacterial action. J. Soc. Chem. Ind. 21: 681-684.

Jung, W.E., and Schweissfurth, R. 1979. Manganese oxidation by an intracellular protein of a *Pseudomonas* species. Z. Allg. Mikrobiol. 19:107-115.

Laddaga, R.A., Chu, L., Misra, T.K., and Silver, S. 1987. Nucleotide sequence and expression of the mercurial resistance operon from *Staphylococcus aureus* plasmid pI258. Proc. Natl, Acad. Sci. USA. 84: 5106-5110.

Larsen, E.I., Sly, L.I., and McEwan, A.G. 1999. Manganese(II) adsorption and oxidation by whole cells and a membrane fraction of *Pedomicrobium* sp. ACM 3067. Arch. Microbiol. 171: 257-264.

Larson, E.J., and Pecoraro, V.L. 1992. Introduction to manganese enzymes. In: Manganese Redox Enzymes. V. L. Pecoraro, ed. VCH Publishers, Inc., New York. p. 1-28.

Lee, Y. 1994. Microbial oxidation of cobalt: characterization and its significance in marine environments. Ph.D. Thesis. University of California, San Diego

Lee, Y., and Tebo, B. M. 1994. Cobalt oxidation by the marine manganese(II)-oxidizing *Bacillus* sp. strain SG-1. Appl. Environ. Microbiol. 60: 2949-2957.

Mathur, A. K., and Dwivedy, K.K. 1988. Biogenic approach to the treatment of uranium mill effluents. Uranium. 4: 385-394.

Matz, L.L., Beaman, T.C., and Gerhardt, P. 1970. Chemical composition of exosporium from spores of *Bacillus cereus*. J. Bacteriol. 101: 196-201.

Mouchet, P. 1992. From conventional to biological removal of iron and manganese in France. J. Am. Water Works. Assoc. 84: 158-162.

Murray, J.W. 1975. The interaction of metal ions at the manganese dioxide-solution interface. Geochim. Cosmochim. Acta. 39: 505-519.

Nealson, K.H., and Ford, J. 1980. Surface enhancement of bacterial manganese oxidation: implications for aquatic environments. Geomicrobiol. J. 2: 21-37.

Nealson, K.H., Rosson, R.A., and Myers, C.R. 1989. Mechanisms of oxidation and reduction of manganese. In: Metal ions and bacteria. T.J. Beveridge and R.J. Doyle, eds. John Wiley and Sons, Inc., New York. p. 383-411.

Nealson, K.H., Tebo, B.M., and Rosson, R.A. 1988. Occurrence and mechanisms of microbial oxidation of manganese. Adv. Appl. Microbiol. 33: 279-318.

Nelson, Y.M., Lion, L.W., Ghiorse, W.C., and Shuler, M.L. 1999. Production of biogenic Mn oxides by *Leptothrix discophora* SS-1 in a chemically defined medium and evaluation of their adsorption characteristics. Appl. Environ. Microbiol. 65: 175-180.

Okazaki, M., Sugita, T., Shimizu, M., Ohode, Y., Iwamoto, K., de Vrind-de Jong, E.W., de Vrind, J.P.M., and Corstjens, P.L.A.M. 1997. Partial purification and characterization of manganese-oxidizing factors of *Pseudomonas fluorescens* GB-1. Appl. Environ. Microbiol. 63: 4793-4799.

Rosson, R.A., and Nealson, K.H. 1982. Manganese binding and oxidation by spores of a marine bacillus. J. Bacteriol. 174: 575-585.

Ryden, L.G., and Hunt, L.T. 1993. Evolution and protein complexity: the blue copper-containing oxidases and related proteins. J. Mol. Evol. 36: 41-66.

Schulze, K., and Roedel, G. 1989. Accumulation of the cytochrome *c* oxidase subunits I and II in yeast requires a mitochondrial membrane-associated protein, encoded by the nuclear SCO1 gene. Mol. Gen. Genet. 216: 37-43.

Solomon, E.I., Sundaram, U.M., and Machonkin, T.E. 1996. Multicopper oxidases and oxygenases. Chem. Rev. 96: 2563-2605.

Stearman, R., Yuan, D.S., Yamaguchi-Iwai, Y., Klausner, R.D., and Dancis, A. 1996. A permease-oxidase complex involved in high-affinity iron uptake in yeast. Science. 271: 1552-1557.

Tebo, B.M. 1995. Metal precipitation by marine bacteria: potential for biotechnological. In: Genetic Engineering-Principles and Methods. J.K. Setlow, ed. Plenum Press, New York. 17: 231-263.

Tebo, B.M. 1998. Comment on "Comment on 'Oxidation of cobalt and manganese in seawater via a common microbially catalyzed pathway' by J.W. Moffett and J. Ho". Geochim. Cosmochim. Acta. 62: 357-358.

Tebo, B.M. 1983. The ecology and ultrastructure of marine manganese oxidizing bacteria. Ph.D. Thesis. University of California, San Diego.

Tebo, B.M., Ghiorse, W.C., van Waasbergen, L.G., Siering, P.L, and Caspi, R. 1997. Bacterially-mediated mineral formation: Insights into manganese(II) oxidation from molecular genetic and biochemical studies. Reviews in Mineralogy. 35: 225-266.

Tebo, B.M., and Lee, Y. 1993. Microbial oxidation of cobalt. In: Biohydrometallurgical Technologies. A. E. Torma, J. E. Wey, and V.L. Lakshmanan, eds. The Minerals, Metals, and Materials Society, Warrendale, PA. p. 695-704.

Tebo, B.M., Mandernack, K., and Rosson, R.A. 1988. Manganese oxidation by a spore coat or exosporium protein from spores of a manganese(II) oxidizing marine bacillus. In: Abstracts of the 88th Annual Meeting of the American Society for Microbiology 1988. American Society for Microbiology, Washington, D. C. p. 201, I-121.

Tebo, B.M., van Waasbergen, L.G., Francis, C.A., He, L.M., Edwards, D.B., and Casciotti, K. 1998. Manganese oxidation by spores of the marine *Bacillus* sp. strain SG-1: application for the bioremediation of metal pollution. In: New Developments in Marine Biotechnology. H.O. Halvorson and Y. Le

Gal, eds. Plenum Press, New York. p. 177-180.

Tipper, D.J., and Gauthier, J.J. 1972. Structure of the bacterial endospore. In: Spores V. L. L. Campbell, ed. American Society for Microbiology, Bethesda, MD. p. 5-12.

van Waasbergen, L.G., Hoch, J.A., and Tebo, B.M. 1993. Genetic analysis of the marine manganese-oxidizing *Bacillus* sp. strain SG-1: protoplast transformation, Tn*917* mutagenesis, and identification of chromosomal loci involved in manganese oxidation. J. Bacteriol. 175: 7594-7603.

van Waasbergen, L.G., Hildebrand, M., and Tebo, B.M. 1996. Identification and characterization of a gene cluster involved in manganese oxidation by spores of the marine *Bacillus* sp. strain SG-1. J. Bacteriol. 12: 3517-3530.

Vojak, P.W.L., Edwards, C., and Jones, M.V. 1984. Manganese oxidation and sporulation by an estuarine *Bacillus* species. Microbios. 41: 39-47.

Warth, A.D. 1978. Molecular structure of the bacterial spore. Adv. Microb. Physiol. 17:1-45.

11

Magnetosome Formation in Magnetotactic Bacteria

Dirk Schüler

Max-Planck-Institut f. Marine Mikrobiologie,
Celsiusstr. 1, D-28359 Bremen, Germany

Abstract

The ability of magnetotactic bacteria to orient and migrate along geomagnetic field lines is based on intracellular magnetic structures, the magnetosomes, which comprise nano-sized, membrane bound crystals of magnetic iron minerals. The formation of magnetosomes is achieved by a biological mechanism that controls the accumulation of iron and the biomineralization of magnetic crystals with a characteristic size and morphology within membrane vesicles. This paper focuses on the current knowledge about magnetotactic bacteria and will outline aspects of the physiology and molecular biology of magnetosome formation. The biotechnological potential of the biomineralization process is discussed.

Introduction

The serendipitous observation of certain mud bacteria, whose swimming direction could be manipulated by magnetic fields, led to the discovery of unique intracellular structures of small magnetic particles known as magnetosomes (Blakemore, 1975). The magnetosomes, which were identified as membrane-bound particles of a magnetic iron mineral (Balkwill et al., 1980), enable the bacteria to orient themselves and swim along the lines of a magnetic field, a behavior referred to as magnetotaxis (Blakemore and Frankel, 1981). The formation of well-ordered, nano-sized particles with perfect magnetic and crystalline properties in magnetotactic bacteria is an intriguing example for a highly controlled biomineralization process, which stimulated an interdisciplinary research following Blakemore's discovery. The mechanisms of magnetosome formation might be relevant for the synthesis of advanced biomaterials with designed properties (Mann, 1993). Commercial uses of bacterial magnetosome particles have been suggested including the manufacture of magnetic tapes and printing inks, magnetic targeting of pharmaceuticals, cell separation and the application as contrast enhancement agents in magnetic resonance imaging (Mann *et al.*, 1990a; Matsunaga, 1991). Recently, ultrafine-grained magnetite particles from a Martian meteorite, which resembled the magnetosome crystals of recent bacteria, have been cited as putative evidence for ancient extraterrestrial life (Frankel *et al.*, 1998; McKay *et al.*, 1996). Moreover, bacterial magnetosome formation might serve as a model system for the biomineralization of magnetic minerals in other organisms, as similar crystals of ferromagnetic material, mainly magnetite, has been found in a wide range of higher organisms and even humans (Kirschvink, 1989; Kirschvink *et al.*, 1992). Thus, an understanding of the structures and mechanisms involved in bacterial biomineralization of magnetosomes is of crucial interest.

This paper will focus on the current knowledge of the ecology, phylogeny and physiology of magnetotactic bacteria from marine and freshwater habitats. Since most of what we know about the mechanism of magnetosome formation comes from studies of *Magnetospirillum* species, emphasis will be given on the biomineralization of magnetite in these bacteria.

Ecology, Occurrence, and Phylogeny of Magnetotactic Bacteria

Magnetotactic bacteria (MTB) are a heterogeneous group of procaryotes which are ubiquitous in aquatic environments and cosmopolitan in distribution. Given their high abundance and variety in marine and freshwater habitats, MTB very likely play an important ecological role in many sediments, as for instance in biogeochemical cycling of iron and other elements. However, their role remains to be fully evaluated. Furthermore, magnetosome particles remain preserved after the bacterial cells die and thus can be deposited as magnetofossils, which significantly contribute to the magnetization of sediments (Petersen *et al.*, 1986; Stolz, 1990).

Freshwater Habitats

Freshwater communities of MTB have been studied by several microscopic, cultural and molecular-biological approaches. Highest numbers of MTB (10^5 -10^6/ml; Blakemore *et al.*, 1979; Spring *et al.*, 1993) are usually found at the oxic-anoxic transition zone generally located at the sediment-water interface, which is consistent with their microaerophilic to anaerobic lifestyle. Freshwater sediments were found to contain various morphological types of MTB, including rod-shaped, vibrio-like, coccoid, and helical forms (Figure 1). With the apparent exception of extreme environments like acidic mine tailings and thermal springs, MTB occur ubiquitously in a wide range of different habitats (Blakemore, 1982; Mann *et al.*, 1990a), and have been even reported from water-logged soils (Fassbinder *et al.*, 1990). Despite their ubiquitous occurrence, different environments appear to support the development of specific populations of MTB. For example, in a survey of several freshwater habitats, the population of MTB in an eutrophic pond was found to contain only three different morphological forms dominated by a species of a magnetic coccus occurring in high numbers (Schüler, 1994). MTB of this type were usually abundant in habitats with high content of organic nutrients (Moench and Konetzka, 1978; Mann *et al.*, 1990a). In contrast, MTB from an

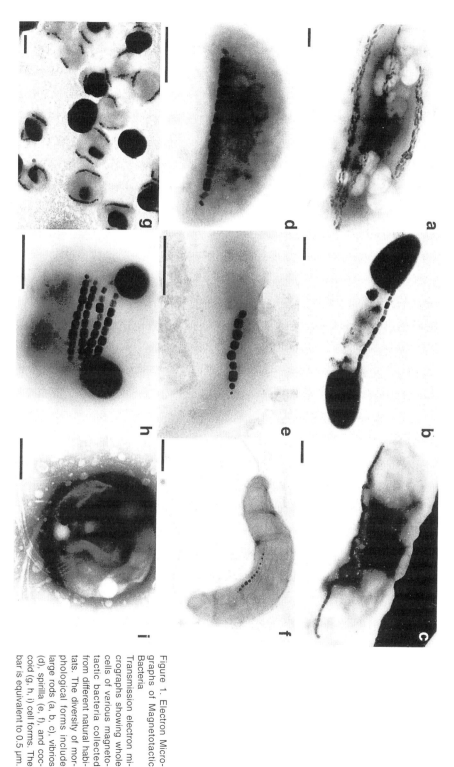

Figure 1. Electron Micrographs of Magnetotactic Bacteria. Transmission electron micrographs showing whole cells of various magnetotactic bacteria collected from different natural habitats. The diversity of morphological forms include large rods (a, b, c), vibrios (d), spirilla (e, f), and coccoid (g, h, i) cell forms. The bar is equivalent to 0.5 µm.

oligotrophic lake sediment were phylogenetically and morphologically more diverse with at least 10 morphologically distinct forms of MTB (Schüler, 1994). In certain microhabitats, cells of a single rod-like species of MTB were the dominant fraction of the microbial community and accounted for up to 30% of the biovolume (Spring et al., 1993). The occurrence of this unusually large bacterium up to 20 µm long and with more then 1000 magnetosome particles, seems to be restricted to certain types of freshwater sediments (Spring et al., 1993). In contrast to what might be anticipated from their remarkable potential to sequester large amounts of iron, habitats of MTB are typically characterized by low to moderate contents of iron (0.01-1 mg/l) and addition of iron does not increase their enrichment (Blakemore, 1982; Oberhack and Süßmuth, 1986). This implies that MTB are able to accumulate iron against a large concentration gradient.

Marine Habitats
In the marine environment, MTB have been mostly found in coastal environments like marshes and estuaries, although some studies are indicating their apparently wide-spread occurrence in the ocean up to depths of 3000 m (Petermann and Bleil, 1993; Stolz et al., 1986). Like in freshwater environments, the occurrence of MTB is usually restricted to the upper layer of the sediment. However, in some chemically stratified estuarine basins, MTB were found to occur in the microaerobic layer of the water-columns (Bazylinski, 1995). Generally, morphological types similar to freshwater MTB are found in marine habitats, although some sulfidic sediments are characterized by the presence of unique morphological forms, notably a many-celled magnetotactic prokaryote which was identified in marine and brackish sediments (Farina et al., 1990; Mann et al. 1990b). Several MTB from reducing environments with high concentrations of H_2S contain particles of iron-sulfide (greigite and pyrite) instead of iron oxides found in most MTB (Bazylinski et al., 1993; Posfai et al., 1998; Mann et al., 1990b). It was demonstrated that a freshwater sulfate-reducing bacterium could intracellularly form magnetite (Sakaguchi et al., 1993), suggesting that marine counterparts are most likely abundant given the importance of sulfate reduction in marine systems.

Phylogeny
Examinations of natural communities by molecular phylogenetic techniques revealed that the morphological diversity of MTB is matched by a remarkable phylogenetic diversity and there is likely a large number of different magnetotactic species. MTB occur in different phylogenetic groups of bacteria, so it has been argued that the ability to form magnetosomes has been evolved from separate evolutionary origins (DeLong et al., 1993). MTB from marine habitats are generally found in the same phylogenetic groups as their freshwater counterparts, although they apparently form slightly separate lineages. So far, there are no known examples of Archaea or Gram-positive bacteria capable of forming magnetosomes. The majority of magnetotactic bacteria including cocci, rods, as well as cultivatable vibrios and spirilla, are members of the α-Proteobacteria (DeLong et al., 1993; Schleifer et al., 1991; Spring et al., 1992; 1994; 1998). Interestingly, based

on comparative sequence analysis of 16S ribosomal RNA, some MTB from the α-subclass are closely related to the nonmagnetic, photosynthetic, non-sulfur purple bacteria, with whom they share the feature of intracytoplasmatic membrane formation. A morphologically distinct large magnetic rod was assigned to the *Nitrospira* phylum, whereas a magnetotactic, many-celled procaryote and a magnetic sulfate-reducing bacterium were found to belong to the δ-subclass of *Proteobacteria* (De Long *et al.*, 1993; Kawaguchi, 1995; Spring and Schleifer, 1995).

Isolation, Cultivation, and Physiology of MTB

Despite their ubiquitous occurrence and high abundance, cultivation of MTB in the laboratory has proven difficult. Problems in isolation and cultivation of these bacteria arise from their lifestyle, which is adapted to sediments and chemically stratified aquatic habitats. As typical gradient organisms, MTB depend on complex patterns of vertical chemical and redox gradients, which are difficult to mimic under laboratory conditions. Since no strictly selective media and growth conditions are known for the cultivation of MTB, the effective separation of magnetotactic cells from non-magnetic contaminants is crucial in their isolation. This can be achieved by exploiting their active migration along magnetic field lines in a capillary "racetrack" method (Wolfe *et al.*, 1987). Growth media involving sulfide and redox gradients have proven useful in the isolation of MTB (Schüler *et al.*, 1999).

   Only a limited number of MTB has been isolated in pure culture so far and most of the isolates are poorly characterized in terms of growth conditions and physiology. Examples of isolates which can be grown under laboratory conditions include several freshwater species of *Magnetospirillum* (Blakemore *et al.*, 1979; Burgess *et al.*, 1993; Schleifer *et al.*, 1991; Schüler and Köhler, 1992; Schüler *et al.*, 1999). Two strains of a marine magnetic vibrio were isolated that are facultative anaerobes and can use either oxygen or nitrous oxide as terminal electron acceptors (Bazylinski *et al.*, 1988; Meldrum *et al.*, 1993a). The only cultivatable magnetic coccus was grown microaerobically in gradient cultures (Meldrum *et al.*, 1993b). Sakaguchi *et al.* (1993) isolated an obligate anaerobic, sulfate-reducing magnetotactic bacterium.

Magnetosomes

The hallmark feature of all MTB is the presence of unique intracellular structures, known as magnetosomes, which consist of magnetic iron mineral particles enclosed within membrane vesicles (Balkwill *et al.*, 1980). With the exception of the aforementioned MTB producing iron-sulfide crystals, the iron mineral particles generally consist of magnetite (Fe_3O_4). Magnetite is an inverse spinel ferrite of structural formula Fe^{3+} [Fe^{2+}, Fe^{3+}]O_4, which has ferrimagnetic properties (Banerjee and Moskowitz, 1985). Unlike magnetite found in inorganic systems or produced extracellularly by the metabolic activities of dissimilatory iron-reducing bacteria (Moskowitz *et al.*, 1989), the intracellular magnetosome crystals are characterized by narrow size distributions and uniform, species-specific crystal habits (Figure 2). All habits can be derived from various combinations of the octahedral {111}, dodecahedral {110}, and cubic {100} forms compatible with magnetite (Fd3m) symmetry (Devouard et al., 1998). The particle sizes are typically 35-120 nm, which is within the permanent, single-magnetic-domain-size range for magnetite (Moskowitz, 1995). The magnetosomes are usually organized in chains, resulting in a permanent magnetic dipole sufficiently large so that it will orient the entire bacterium along the geomagnetic field at ambient temperature. This passive orientation results in the migration of the cell along the magnetic field lines as it swims ("Magnetotaxis"; Blakemore and Frankel, 1981).

Function of Magnetotaxis

Because all known magnetotactic bacteria are either microaerophilic or anaerobic, a widely accepted hypothesis regarding the function of magnetotaxis is that they seek to avoid high, potentially toxic oxygen tension. That is, their navigation along the geomagnetic field lines facilitates migration to their favored position in the oxygen gradient (Blakemore and Frankel, 1981). The preferred motility direction found in natural populations of magnetotactic bacteria is northward in the geomagnetic field in the northern hemisphere, whereas it is southward in the southern hemisphere. Because of the inclination of the geomagnetic field, migration in these preferred directions would cause cells in both hemispheres to swim downward. Recent findings indicate that this process is more complex and apparently involves interactions with an aerotactic sensory mechanism (Frankel et al., 1997). Besides magnetotaxis, other possible explanations for the intracellular iron deposition have been considered, including functions in iron homeostasis, energy conservation, or redox cycling (Mann et al., 1990a; Guerin and Blakemore, 1992; Spring et al., 1993).

Physiology, Biochemistry, and Molecular Biology of Magnetite Biomineralization in Members of the Genus *Magnetospirillum*

Most of our knowledge of the mechanism of magnetite biomineralization comes from studies involving several strains of the genus *Magnetospirillum*, although recently efforts were made to address magnetosome synthesis in other systems (Dubbels et al., 1998; Dean and Bazylinski, 1999). Strains of *Magnetospirillum* can be grown microaerobically on simple media containing short organic acids as a carbon source (Blakemore et al., 1979; Matsunaga et al., 1990; Schleifer et al., 1991; Schüler and Köhler, 1992) and have a oxygen-dependent respiratory type of metabolism, although nitrate can be used microaerobically as an electron acceptor in some strains (Bazylinski and Blakemore, 1983). *M. magnetotacticum* was the first magnetotactic bacterium available in pure culture and was used in initial studies. One of the reasons for oxygen sensitivity in this organism may be the lack of the oxygen-protective enzyme catalase, as addition of catalase to the growth medium resulted in increased oxygen tolerance (Blakemore et al., 1979). *M. gryphiswaldense* and *Magnetospirillum* AMB-1 are more oxygen tolerant, which obviates the needs of elaborate microaerobic techniques and facilitates

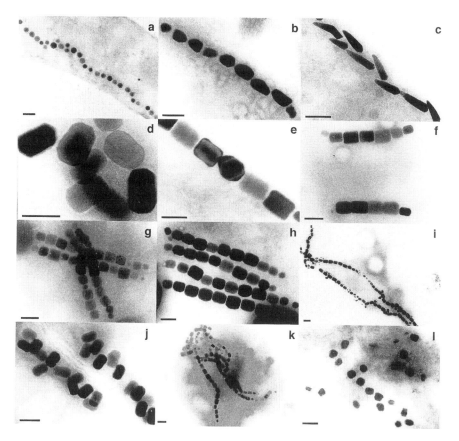

Figure 2. Electron Micrographs of Magnetosomes
Crystal morphologies and intracellular organization of magnetosomes found in various magnetotactic bacteria. Shapes of magnetic crystals include cubo-octahedral (a), bullet-shaped (b, c) elongated prismatic (d, e, f, g, h, i, j, k) and rectangular morphologies (l). The magnetosome particles can be arranged in one (a, b, c, e), two (f, i) or multiple chains (g, h) or irregularly (j, k, l). The bar is equivalent to 0.1 μm.

the mass cultivation of magnetic cells (Schüler and Baeuerlein, 1997a; Matsunaga *et al.*, 1990).

Cells of *Magnetospirillum* species form cubo-octahedral magnetite crystals, which are 42-45 nm in size and arranged in a chain (Figure 2a). On the basis of high resolution electron microscopy, Mößbauer spectroscopy and biochemical results, a model for magnetite biomineralization was proposed (Frankel *et al.*, 1983; Mann *et al.* 1990a), in which $Fe(III)$ is taken up by the cell, possibly via a reductive step (Paoletti and Blakemore, 1988). Iron is then thought to be reoxidized to form a low-density hydrous oxide which is dehydrated to form a high-density $Fe(III)$ oxide (ferrihydrite). In the last step, one-third of the $Fe(III)$ ions are reduced, and with further dehydration, magnetite is produced within the magnetosome vesicle (Figure 3).

Because of the large amounts of iron required for magnetite synthesis, magnetotactic bacteria can be anticipated to use very efficient uptake systems for the assimilation of iron. Although several studies have focused on iron

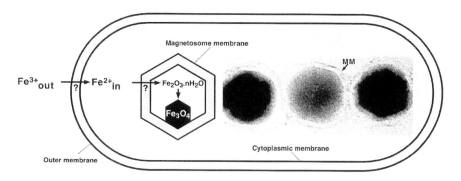

Figure 3 Proposed Model for Magnetite Biomineralization in *Magnetospirillum* species
Fe(III) is actively taken up by the cell, possibly via a reductive step. Iron is then thought to be reoxidized to form a low-density hydrous oxide which is dehydrated to form a high-density Fe(III) oxide (ferrihydrite). In the last step, one-third of the Fe(III) ions are reduced, and with further dehydration, magnetite is produced within the magnetosome vesicle. The magnetosome membrane contains specific proteins, which are thought to have crucial functions in the accumulation of iron, nucleation of minerals and redox and pH control. The electron micrograph shows magnetosome particles from *Magnetospirillum gryphiswaldense*.

uptake, the exact mechanisms and components involved are not well understood. The initial finding of Paoletti and Blakemore (1986) that cells of *M. magnetotacticum* produced a hydroxamate type siderophore under high iron condition, has not been replicated in other studies and no unequivocal evidence for the involvement of siderophores in the formation of magnetite has been found so far. Nakamura *et al.* (1993a) hypothesized that ferric iron was taken up in *Magnetospirillum* AMB-1 by a periplasmic binding protein-dependent iron transport system. In *M. gryphiswaldense*, the major portion of iron is taken up as Fe(III) in an energy-dependent process with a V_{max} and K_m of 0.86 nmol min^{-1} (mg dry weight)$^{-1}$ and 3 µM, respectively (Schüler and Baeuerlein, 1996). The high rates of ferric iron uptake may reflect the extraordinary requirement for iron in these bacteria. Both the amount of magnetite formed and the rates of iron uptake were close to maximum at extracellular iron concentrations of 15-20 µM Fe, indicating that this bacterium is able to accumulate copious amounts of iron from relatively low concentrations.

Indirect evidence suggests that ferritin-like iron storage proteins may be involved in the biomineralization of magnetite. Magnetite in the radula of Polylplacophoran mollusks is always formed by conversion of pre-existing ferritin (Kirschvink and Lowenstam, 1979). Studies of *M. magnetotacticum* have revealed the presence of a high-density hydrous ferric oxide (probably ferritin) within intact cells (Frankel *et al.*, 1983). Hence, it was concluded that these bacteria may use a biochemical pathway in magnetite formation similar to that of the Polyplacophoran mollusks. Recently, evidence for the presence of two overlapping bacterioferritin genes was found in *M. magnetotacticum* (Bertani *et al.*, 1997), but its relevance for magnetite formation requires further study.

While size and morphologies of mature magnetite crystals are largely unaffected by environmental conditions, the number of magnetosome particles per cell can vary considerably and strongly depends on the growth conditions. Besides the availability of micromolar amounts of iron, microaerobic conditions are required for magnetite formation in *Magnetospirillum* species (Blakemore *et al.*, 1985; Schüler *et al.*, 1995; Schüler and Baeuerlein, 1998). Cells of *M. gryphiswaldense*, which are non-magnetic during aerobic growth, start to produce Fe_3O_4 when shifted to microaerobic growth conditions corresponding to an oxygen concentration of about 2-7 μM O_2. The accumulation of iron during growth is tightly coupled to the induction of magnetite biomineralization. While the intracellular iron content in pre-magnetic cells is relatively low and in the same range as reported for other, non-magnetotactic bacteria, magnetic spirilla can accumulate more then 2 % iron on a dry weight basis (Blakemore *et al.*, 1979; Schüler and Baeuerlein, 1998).

In an attempt to understand the relationship between respiratory electron transport in nitrate and oxygen utilization and magnetite synthesis, cytochromes in *M. magnetotacticum* have been examined. Tamegai *et al.* (1993) reported a novel cytochrome a_1-like hemoprotein that was present in greater amounts in magnetic cells than nonmagnetic cells. A new ccb-type cytochrome c oxidase (Tamegai and Fukumori, 1994) and a cytochrome cd_1-type nitrite reductase (Yamazaki *et al.*, 1995) were isolated and purified from the same organism. The latter protein is of particular interest since it showed Fe(II):nitrite oxido-reductase activity. It was proposed by Fukumori *et al.* (1997) that the dissimilatory nitrite reductase of *M. magnetotacticum* could function as Fe(II) oxidizing enzyme for magnetite synthesis under anaerobic conditions, which is consistent with the observation that magnetite formation is enhanced in microaerobically denitrifying cells of *M. gryphiswaldense* (Blakemore *et al.*, 1985). The purification of a ferric iron reductase which is loosely bound to the cytoplasmic membrane was described in *M. magnetotacticum* (Noguchi *et al.*, 1999). Based on the observation that the iron reductase activity was correlated with the magnetosome number in the cells if grown in the presence of different concentrations of Zn^{2+}, a participation of the enzyme in magnetite synthesis was suggested.

In strains of *Magnetospirillum*, the magnetosome mineral phase is enveloped by a trilaminate membrane (Gorby *et al.*, 1988; Matsunaga, 1991; Schüler and Baeuerlein, 1997b). Compartmentalization through the formation of the magnetosome vesicle enables the processes of mineral formation to be regulated by biochemical pathways. The membrane may act as potential gate for compositional, pH and redox differentiation between the vesicle and the cellular environment. Although the magnetosome membrane (MM) does not appear to be contiguous with the cell membrane in electron-microscopic images, it can be speculated that some sort of physical connection with the cytoplasmic membrane exists, which could explain the biosynthetic origin of the magnetosome compartment and the anchoring of the magnetosome chain in the cell. The MM vesicles apparently exist prior to the biomineralization of the mineral phase, because empty and partially filled vesicles have been observed in iron starved cells of *M. magnetotacticum* and *M. gryphiswaldense* (Gorby *et al.*, 1988; Schüler and Baeuerlein, 1997b). The MM is composed of phospholipids and proteins, at least several of which appear unique to this membrane. Although the protein patterns of the magnetosome membrane are different between several species of

Magnetospirillum, at least one major protein with a molecular weight of about 22-24 kDa appears to be common to all strains tested so far as revealed by sequence analysis and antibody cross-reactivity (Okuda *et al.*, 1996; Schüler *et al.*, 1997; Schüler and Tebo, unpublished). The exact role of the magnetosome-specific proteins has not been elucidated, but it has been suggested that these have specific functions in the accumulation of iron, nucleation of minerals and redox and pH control (Gorby *et al.*, 1988; Mann *et al.*, 1990a).

Attempts to address the molecular genetics of magnetosome formation in MTB have been hampered by several problems including the lack of a significant number of magnetotactic bacterial strains and the elaborate techniques required for the growth of these bacteria. Therefore, our knowledge of the genetic determination of magnetosome formation is still limited. In initial studies, Berson *et al.* (1989) demonstrated that genes (e. g. *recA*) of *M. magnetotacticum* can be functionally expressed in *Escherichia coli*. They also cloned a 2 kb DNA fragment from *M. magnetotacticum* that complemented iron uptake deficiencies in *E. coli* and *Salmonella typhimurium* mutants lacking a functional *aroD* gene (biosynthetic dehydroquinase). This suggests that the 2 kb DNA fragment may have a function in iron uptake in *M. magnetotacticum* (Berson *et al.*, 1991).

Okuda *et al.* (1996) used the N-terminal amino acid sequence from a 22 kDa protein specifically associated with the MM to clone and sequence its respective gene. Based on the amino acid sequence, the protein exhibits significant homology with a number of functionally diverse proteins belonging to the tetratricopeptide repeat family, although its function in magnetosome synthesis is not clear.

Because of their microaerophilic nature, colony formation on plates is not easy to obtain in MTB and the poor ability of most strains to form colonies on plates has been a major obstacle for the establishment of genetic experiments. It has been demonstrated that strains of *Magnetospirillum* and related bacteria are otherwise accessible by genetic manipulation (Matsunaga *et al.*, 1992; Eden and Blakemore, 1991). Transfer of DNA has been achieved by conjugal transfer of broad host range IncP and IncQ plasmids, which formed autonomous replicons in *Magnetospirillum* sp. AMB-1. Cells of this strain were reported to form magnetic colonies on the surface of agar plates when grown under an atmosphere containing 2% oxygen (Matsunaga *et al.*, 1991). This facilitated the selection of non-magnetic mutants generated by Tn5-mutagenesis (Matsunga *et al.*, 1992). It was concluded that at least three regions of the chromosome are required for the synthesis of magnetosomes. One of these regions contained a gene, designated *magA*, encoding a protein with homology to cation efflux proteins, in particular the *E. coli* potassium ion-translocating protein KefC (Nakamura *et al.*, 1995a). Membrane vesicles prepared from *E. coli* cells expressing *magA* took up iron when ATP was supplied indicating that energy was required for the uptake of iron. The expression of *magA* was enhanced when wild-type *Magnetospirillum* AMB-1 cells were grown under iron-limited conditions rather than iron-sufficient conditions in which they would produce magnetosomes (Nakamura *et al.*, 1995b). Thus, the role of the *magA* gene in magnetosome formation requires further clarification.

Biotechnological Applications

Magnetosome formation in bacteria has the potential to yield useful biomaterials. The synthesis of a nano-sized magnetite particles by a biological process promises advantages in terms of controlling crystal growth and structural properties. By comparison, synthetic magnetic particles are non-uniform, often not fully crystalline, compositionally nonhomogenous, and in an agglomerated state which imposes problems in processing (Sarikaya, 1994). Moreover, biomineralization provides a way to produce highly uniform magnetite crystals without the drastic regimes of temperature, pH and pressure which are often needed for their industrial production (Mann *et al.*, 1990a). Accordingly, numerous biotechnological applications of the small magnetic crystals have been contemplated (Schüler and Frankel, 1999).

The small size of isolated magnetosome particles provides a large surface-to-volume ratio, which makes them useful as carriers for the immobilization of relatively large quantities of bioactive substances, which can then be separated by magnetic fields. In several studies, bacterial magnetosomes were used for immobilizing enzymes and antibodies (Matsunaga and Kamyia, 1987; Nakamura *et al.*, 1991; 1993b). The presence of the magnetosome membrane resulted in dispersion and handling properties superior to those of synthetic magnetic particle conjugates (Matsunga, 1991). Using phagocytosis and polyethylene glycol fusions, bacterial magnetite particles were incorporated into eucaryotic cells, which could be manipulated by magnetic fields (Matsunaga *et al.*, 1989).

Another application of bacterial magnetosomes may lie in their potential use as a contrast agent for magnetic resonance imaging and tumor-specific drug carriers based on intratumoral enrichment. Synthetic liposomes containing superparamagnetic iron oxide particles have been already used in biomedical applications of this type (Päuser *et al.*, 1997). Other suggested applications of MTB involve the use of living, actively swimming cells. Proposed examples include the application of MTB in the analysis of magnetic domains in magnetic materials (Harasko *et al.*, 1993; 1995) and the use of MTB for the removal of heavy metals and radionuclides from wastewater (Bahaj *et al.*, 1994; 1998).

As yet, no application of MTB has attained commercial scale, however. This situation is in part due to the problems related to mass cultivation of these bacteria. Another reason for the limited practicability of many systems is the lack of an basic understanding of bacterial magnetite biomineralization at a biochemical and molecular level. Thus, more research in this field is clearly required. Genetic approaches are offering promising ways for engineering the biomineralization process and providing new materials. Questions in further research could be addressed in relatively well-studied systems such as members of the genus *Magnetospirillum*, while at the same time increased effort needs to be spent to exploit the metabolic potential of the vast biodiversity of natural occurring MTB, including largely unexplored habitats such as the marine environment.

Acknowledgements

This work was supported by grants from the Deutsche Forschungsgemeinschaft and the Max-Planck-Society. I am grateful to E. Baeuerlein, D.A. Bazylinski, R.B. Frankel, M. Haygood, M. Hildebrand, S. Spring and B.M. Tebo for discussions and collaboration, and D.L. Parker for valuable suggestions regarding the manuscript.

References

Bahaj, A.S., Croudace, I.W., James, P.A.B., Moeschler, F.D., and Warwick, P.E. 1998. Continous radionuklide recovery from wastewater using magnetotactic bacteria. J. Inorg. Biochem. 59: 107.

Bahaj, A.S., Croudace, I.W., and James, P.A.B. 1994. Treatment of heavy metal contaminants using magnetotactic bacteria. IEEE Transactions on Magnetics 30: 4707-4709.

Balkwill, D., Maratea, D., and Blakemore, R.P. 1980. Ultrastructure of a magnetotactic spirillum. J. Bacteriol. 141: 1399-1408.

Banerjee, S. K., and Moskowitz, B. M. 1985. Ferrimagnetic properties of magnetite. In: Magnetite Biomineralization and Magnetoreception in Organisms. J.L. Kirschvink, D.S. Jones, B.M. MacFadden, eds. Plenum Press, New York and London. p. 17-41.

Bazylinski, D.A. 1995. Structure and function of the bacterial magnetosome. ASM News. 61: 337-343.

Bazylinski, D.A., and Blakemore, R.P. 1983. Denitrification and assimilatory nitrate reduction in *Aquaspirillum magnetotacticum*. Appl. Environ. Microbiol. 46: 1118-1124.

Bazylinski, D.A., Frankel, R.B., and Jannasch, H.W. 1988. Anaerobic magnetite production by a marine magnetotactic bacterium. Nature. 334: 518-519.

Bazylinski, D.A., Heywood, B.R., Mann, S., and Frankel, R.B. 1993. Fe_3O_4 and FeS_4 in a bacterium. Nature. 366: 218.

Bazylinski, D.A., Soohoo, C.K., and Hollocher, T.C. 1986. Growth of Pseudomonas aeruginosa on nitrous oxide. Appl. Environ. Microbiol. 51: 1239-1246.

Berson, A.E., Hudson, D.V., and Waleh, N.S. 1989. Cloning and characterization of the *recA* gene of *Aquaspirillum magnetotacticum*. Arch. Microbiol. 152: 567-571.

Berson, D.A., Hudson, D.V., and Waleh, N.S. 1991. Cloning of a sequence of *Aquaspirillum magnetotacticum* that complements the *aroD* gene of *Escherichia coli*. Mol. Microbiol. 5: 2262-2264.

Bertani, L.E., Huang, J.S., Weir, B.A., and Kirschvink, J.L. 1997. Evidence for two types of subunits in the bacteriferritin of *Magnetospirillum magnetotacticum*. Gene. 201: 31-36.

Blakemore, R.P. 1975. Magnetotactic bacteria. Science 190: 377-379.

Blakemore, R.P. 1982. Magnetotactic bacteria. Ann. Rev. Microbiol. 36: 217-238.

Blakemore, R.P., and Frankel, R.B. 1981. Magnetic navigation in bacteria. Scientific American. 245: 42-49.

Blakemore, R.P., Maratea, D., and Wolfe, R.S. 1979. Isolation and pure culture of a freshwater magnetic spirillum in chemically defined medium. J. Bacteriol. 140: 720-729.

Blakemore, R.P., Short, K., Bazylinski, D., Rosenblatt, C., and Frankel, R.B. 1985. Microaerobic conditions are required for magnetite formation within *Aquaspirillum magnetotacticum*. Geomicrobiol. J. 4: 53-71.

Burgess, J.G., Kawaguchi, R., Sakaguchi, T., Thornhil, R.H., and Matsunaga, T. 1993. Evolutionary relationships among *Magnetospirillum* strains inferred from phylogenetic analysis of 16S-rRNA sequences. J. Bacteriol. 175: 6689-6694.

Dean, A.J., and Bazylinski, D.A. 1999. Genome analysis of several marine, magnetotactic bacterial strains by pulsed-field gel electrophoresis. Curr. Microbiol. 39: 219-225.

DeLong, E.F., Frankel, R.B., and Bazylinski, D.A. 1993. Multiple evolutionary origins of magnetotaxis in bacteria. Science. 259: 803-806.

Devouard, B., Posfai, M., Hua, X., Bazylinski, D.A., Frankel, R.B., and Buseck, P.R. 1998. Magnetite from magnetotactic bacteria: Size distributions and twinning. American Mineralogist. 83: 1387-1398.

Dubbels, B.L., Dean, A.J., and Bazylinski, D.A. 1998. Approaches to and studies in understanding the molecular basis for magnetosome synthesis in magnetotactic bacteria. Abstracts of the general meeting of the ASM, Atlanta, USA. 290.

Eden, P.A., and Blakemore, R.P. 1991. Electroporation and conjugal plasmid transfer to members of the genus *Aquaspirillum*. Arch. Microbiol. 155: 449-452.

Farina, M., Esquivel, D.M.S., and Lins de Barros, H.G.P. 1990. Magnetic iron-sulphur crystals from a magnetotactic microorganism. Nature. 343: 257-259.

Fassbinder, J.W.E., Stanjek, H., and Vali, H. 1990. Occurence of magnetic bacteria in soil. Nature. 343: 161-163.

Frankel, R.B., Bazylinski, D.A., and Schüler, D. 1998. Biomineralization of magnetic iron minerals in magnetotactic bacteria. Supramolecular Science. 5: 383-390.

Frankel, R.B., Bazylinski, D.A., Johnson, M.S., and Taylor, B.L. 1997. Magneto-aerotaxis in marine coccoid bacteria. Biophys. J. 73: 994-1000.

Frankel, R.B., Papaefthymiou, G.C., Blakemore, R.P., and O'Brien, W. 1983. Fe_3O_4 precipitation in magnetotactic bacteria. Biochim. Biophys. Acta 763: 147-159.

Fukumori, Y., Oynagi, H., Yoshimatsu, K., Noguchi, Y., and Fujiwara, T. 1997. Enzymatic iron oxidation and reduction in magnetite synthesizing *Magnetospirillum magnetotacticum*. J. Phys. IV 7: 659-662.

Gorby, Y.A., Beveridge, T.J., and Blakemore, R.P. 1988. Characterization of the bacterial magnetosome membrane. J. Bacteriol. 170: 834-841.

Guerin, W.F., and Blakemore, R.P. 1992. Redox cycling of iron supports growth and magnetite synthesis by *Aquaspirillum magnetotacticum*. Appl. Environ. Microbiol. 58: 1102-1109.

Harasko, G., Pfützner, H., Rapp, E., Futschik, K., and Schüler, D. 1993. Determination of the concentration of magnetotactic bacteria by means of susceptibility measurements. Jpn. J. Appl. Phys. 32: 252-260.

Harasko, G., Pfützner, H., and Futschik, K. 1995. Domain analysis by means of magnetotactic bacteria. IEEE Transactions on Magnetics. 31: 938-949.

Kawaguchi, R., Burgess, G.J., Sakaguchi, T., Takeyama, H., Thornhill, R.H., and Matsunaga, T. 1995. Phylogenetic analysis of a novel sulfate-reducing magnetic bacterium, RS-1, demonstrates its membership of the δ-proteobacteria. FEMS Microbiol. Lett. 126: 277-282.

Kirschvink JL, and Lowenstam HA. 1979. Mineralization and magnetization of chiton teeth: Paleomagnetic, sedimentologic, and biologic implications of organic magnetite. Earth Planet. Sci. Lett. 44: 193-204.

Kirschvink, J.L. 1989. Magnetite biomineralization and geomagnetic sensitivity in animals: an update and recommendations for future study. Bioelectromagnetics. 10: 239-259.

Kirschvink, J.L., Kobayashi-Kirschvink, A., and Woodford, B.J. 1992. Magnetite biomineralization in the human brain. Proc. Natl. Acad. Sci. USA. 89: 7683-7687.

Mann, S. 1993. Molecular tectonics in biomineralization and biomimetic materials chemistry. Nature. 365: 499-505.

Mann, S., Sparks, N.H.C., and Board, R.G. 1990a. Magnetotactic bacteria: microbiology, biomineralization, palaeomagnetism and biotechnology. Adv. Microbiol. Physiol. 31: 125-181.

Mann, S., Sparks, N.H.C., Frankel, R.B., Bazylinski, D.A., and Jannasch, H.W. 1990b. Biomineralization of ferrimagnetic greigite (Fe_3O_4) and iron pyrite (FeS_2) in a magnetotactic bacterium. Nature. 343: 258-260.

Matsunaga, T. 1991. Applications of bacterial magnets. Trends Biotechnol. 9: 91-95.

Matsunaga, T., Hashimoto, K., Nakamura, N., Nakamura, K., and Hashimoto, S. 1989. Phagocytosis of bacterial magnetite by leukocytes. Appl. Microbiol. Technol. 31: 401-405.

Matsunaga, T., and Kamiya, S. 1987. Use of magnetic particles isolated from magnetotactic bacteria for enzyme immobilization. Appl. Microbiol. Biotechnol. 26: 328-332.

Matsunaga, T., Nakamura, C., Burgess, J.G., and Sode, S. 1992. Gene transfer in magnetic bacteria: Transposon mutagenesis and cloning of genomic DNA fragments required for magnetosome synthesis. J. Bacteriol. 174: 2748-2753.

Matsunaga, T., Sakaguchi, T., and Tadokoro, F. 1991. Magnetite formation by a magnetic bacterium capable of growing aerobically. Appl. Microbiol. Biotechnol. 35: 651-655.

Matsunaga, T., Tadokoro, F., and Nakamura, N. 1990. Mass culture of magnetic bacteria and their application to flow type immunoassays. IEEE Trans. Magnet. Mag. 26: 1557-1559.

McKay, D.S., Gibson, E.K.J., Thomas-Keprta, K.L., Vali, H., Romanek, C.S., Clemett, S.J., Chillier, X.D.F., Maechling, C.R., and Zare, R.N. 1996. Search for past life on Mars: Possible relic biogenic activity in Martian meteorite ALH84001. Science. 273: 925-867.

Meldrum, F.C., Mann, S., Heywood, B.R., Frankel, R.B., and Bazylinski, D.A. 1993a. Electron microscopy study of magnetosomes in two cultured vibroid magnetotacic bacteria. Proc. R. Soc. Lond. B 251: 237-242.

Meldrum, F.C., Mann, S., Heywood, B.R., Frankel, R.B., and Bazylinski, D.A.

1993b. Electron microscopy study of magnetosomes in a cultured coccoid magnetotactic bacterium. Proc. R. Soc. Lond. B 251: 231-236.

Moench, T.T., and Konetzka, W.A. 1978. A novel method for the isolation and study of a magnetotactic bacterium. Arch. Microbiol. 119: 203-212.

Moskowitz, B.M. 1995. Biomineralization of magnetic minerals. Rev. of Geophysics. 33: 123-128.

Moskowitz, B.M., Frankel, R.B., Bazylinski, D.A., Jannasch, H.W., and Lovley, D.R. 1989. A comparison of magnetite particles produced anaerobically by magnetotactic and dissimilatory iron-reducing bacteria. Geophys. Res. Lett. 16: 665-668.

Nakamura, C.B., Burgess, J.G., Sode, K., and Matsunaga, T. 1995a. An iron-regulated gene, *magA*, encoding an iron transport protein of *Magnetospirillum* sp. strain AMB-1. J. of Biol. Chem. 270: 28392-28396.

Nakamura, C., Kikuchi, T., Burgess, J.G., and Matsunaga, T. 1995b. Iron-regulated expression and membrane localization of the magA protein in *Magnetospirillum* sp. strain AMB-1. J. Biochem. 118: 23-27.

Nakamura, C., Sakaguchi, T., Kudo, S., Burgess, J.G., Sode, K., and Matsunaga, T. 1993a. Characterization of iron uptake in the magnetic bacterium *Aquaspirillum* sp. AMB-1. Appl. Biochem. Biotechnol. 39/40: 169-177.

Nakamura, N., Burgess, J.G., Yagiuda, K., Kiudo, S., Sakaguchi, T., and Matsunaga, T. 1993b. Detection and removal of *Escherichia coli* using fluorescin isothiocyanate conjugated monoclonal antibody immobilized on bacterial magnetic particles. Anal. Chem. 65: 2036-2039.

Nakamura, N., Hashimoto, K., and Matsunaga, T. 1991. Immunoassay method for the determination of immunoglobulin G using bacterial magnetic particles. Anal. Chem. 63: 268-272.

Noguchi, Y., Fujiwara, T., Yoshimatsu, K., and Fukumori, Y. 1999. Iron reductase for magnetite synthesis in the magnetotactic bacterium *Magnetospirillum magnetotacticum*. J. Bacteriol. 181: 2142-2147.

Oberhack, M., and Süßmuth, R. 1986. Magnetotactic bacteria from freshwater. Z. Naturforsch. 42c: 300-306.

Okuda, Y., Denda, K., and Fukumori, Y. 1996. Cloning and sequencing of a gene encoding a new member of the tetratricopeptide protein family from magnetosomes of *Magnetospirillum magnetotacticum*. Gene. 171: 99-102.

Paoletti, L.C., and Blakemore, R.P. 1986. Hydroxamate produktion by *Aquaspirillum magnetotacticum*. J. Bacteriol. 167: 153-163.

Paoletti, L.C., and Blakemore, R.P. 1988. Iron reduction by *Aquaspirillum magnetotacticum*. Curr. Microbiol. 17: 339-342.

Päuser, S., Reszka, R., Wagner, S., Wolf, K.J., Buhr, H.J., and Berger, G. 1997. Liposome-encapsulated superparamagnetic iron oxide particles as markers in an MRI-guided search for tumor-specific drug carriers. Anti-Cancer Drug Design. 12: 125-135.

Petermann, H., and Bleil, U. 1993. Detection of live magnetotactic bacteria in south-atlantic deep-sea sediments. Earth Plan. Sci. Lett. 117: 223-228.

Petersen, N., Dobeneck, T.v., and Vali, H. 1986. Fossil bacterial magnetite in deep-sea sediments from the South Atlantic Ocean. Nature. 320: 611-615.

Posfai, M., Buseck, P.R., Bazylinski, D.A., and Frankel, R.B. 1998. Reaction sequence of iron sulfide minerals in bacteria and their use as biomarkers. Science. 280: 880-883.

Sakaguchi, T., Burgess, J.G., and Matsunaga, T. 1993. Magnetite formation by a sulphate-reducing bacterium. Nature. 365: 47-49.

Sarikaya, M. 1994. An introduction to biomimetics: A structural viewpoint. Microsopy Res. Tech. 27: 360-375.

Schleifer, K.H., Schüler, D., Spring, S., Weizenegger, M., Amann, R., Ludwig, W., and Köhler, M. 1991. The genus *Magnetospirillum* gen. nov., description of *Magnetospirillum gryphiswaldense* sp. nov. and transfer of *Aquaspirillum magnetotacticum* to *Magnetospirillum magnetotacticum* comb. nov. System. Appl. Microbiol. 14: 379-385.

Schüler, D. 1994. Isolierung und Charakterisierung magnetischer Bakterien - Untersuchungen zur Magnetitbiomineralisation in *Magnetospirillum gryphiswaldense*. Doctoral Thesis. TU München, München.

Schüler, D., and Baeuerlein, E. 1997a. The biomineralization of magnetite in magnetic bacteria. In: Bioinorganic Chemistry: Transition Metals in Biology and Coordination Chemistry/Deutsche Forschungsgemeinschaft. A. Trautwein, ed. Wiley-VCH, Weinheim; New York; Chichester; Brisbane; Singapore; Toronto. p. 24-36.

Schüler, D., and Baeuerlein, E. 1997b. Iron transport and magnetite crystal formation of the magnetic bacterium *Magnetospirillum gryphiswaldense*. J. Phys. IV 7: 647-650.

Schüler, D., and Baeuerlein, E. 1996. Iron-limited growth and kinetics of iron uptake in *Magnetospirillum gryphiswaldense*. Arch. Microbiol. 166: 301-307.

Schüler, D., and Baeuerlein, E. 1998. Dynamics of iron uptake and Fe_3O_4 biomineralization during aerobic and microaerobic growth of *Magnetospirillum gryphiswaldense*. J. Bacteriol. 180: 159-162.

Schüler, D., Baeuerlein, E., and Bazylinski, D.A. 1997. Localization and occurrence of magnetosome proteins from *Magnetospirillum gryphiswaldense* studied by an immunological method. Abstracts of the general meeting of the ASM. 1997. 332.

Schüler, D., and Frankel, R.B. 1999. Bacterial magnetosomes: Microbiology, biomineralization and biotechnological applications. Appl. Microbiol. Biotechnol. 52: 464-473.

Schüler, D., and Köhler, M. 1992. The isolation of a new magnetic spirillum. Zentralbl. Mikrobiol. 147: 150-151.

Schüler, D., Spring, S., and Bazylinski, D.A. 1999. Improved technique for the isolation of magnetotactic spirilla from a freshwater sediment and their phylogenetic characterization. Systematic Appl. Microbiol. 22: 466-471.

Schüler, D., Uhl, R., and Baeuerlein, E. 1995. A simple light-scattering method to assay magnetism in *Magnetospirillum gryphiswaldense*. FEMS Microbiol. Lett. 132: 139-145.

Spring, S., Amann, R., Ludwig, W., Schleifer, K.H., Gemerden, H.v., and Petersen, N. 1993. Dominating role of an unusual magnetotactic bacterium in the microaerobic zone of a freshwater sediment. Appl. Environ. Microbiol. 59: 2397-2403.

Spring, S., Amann, R., Ludwig, W., Schleifer, K.H., and Petersen, N. 1992. Phylogenetic diversity and identification of nonculturable magnetotactic bacteria. System. Appl. Microbiol. 15: 116-122.

Spring, S., Amann, R., Ludwig, W., Schleifer, K.H., Schüler, D., Poralla, K., and Petersen, N. 1994. Phylogenetic analysis of uncultured magnetotactic bacteria from the alpha-subclass of proteobacteria. System. Appl. Microbiol. 17: 501-508.

Spring, S., Lins, U., Amann, R., Schleifer, K.H., Ferreira, L.C.S., Esquivel, D.M.S., and Farina, M. 1998. Phylogenetic affiliation and ultrastructure of uncultured magnetic bacteria with unusual large magnetosomes. Arch. Microbiol. 169: 136-147.

Spring, S., and Schleifer, K.H. 1995. Diversity of magnetotactic bacteria. System. Appl. Microbiol. 18: 147-153.

Stolz, J.F. 1990. Biogenic magnetite and the magnetization of sediments. J. Geophys. Res. 95: 4355-4361.

Stolz, J.F., Chang, S.B.R., and Kirschvink, J.L. 1986. Magnetotactic bacteria and single-domain magnetite in hemipelagic sediments. Nature. 321: 849-851.

Tamegai, H., and Fukumori, Y. 1994. Purification and some molecular and enzymatic features of a novel ccb-type cytochrome c oxidase from a microaerobic denitrifier, *Magnetospirillum magnetotacticum*. FEBS Lett. 347: 22-26.

Tamegai, H., Yamanaka, T., and Fukumori, Y. 1993. Purification and properties of a ´cytochrom a_1´-like hemoprotein from a magnetotactic bacterium, *Aquaspirillum magnetotacticum*. Biochim. Biophys. Acta 1158: 137-243.

Wolfe, R.S., Thauer, R.K., and Pfennig, N. 1987. A capillary racetrack method for isolation of magnetotactic bacteria. FEMS Microbiol. Lett. 45: 31-35.

Yamazaki, T., Oyanagi, H., Fujiwara, T., and Fukumori, Y. 1995. Nitrite reductase from the magnetotactic bacterium *Magnetospirillum magnetotacticum* - a novel cytochrome-cd(1) with Fe(II)-nitrite oxidoreductase activity. Eur. J. Biochem. 233: 665-671.

12

Pressure Response in Deep-sea Piezophilic Bacteria

Chiaki Kato, Mohammad Hassan Qureshi, and Koki Horikoshi

The DEEPSTAR Group, Japan Marine Science
and Technology Center, 2-15 Natsushima-cho,
Yokosuka 237-0061, Japan

Abstract

Several piezophilic bacteria have been isolated from deep-sea environments under high hydrostatic pressure. Taxonomic studies of the isolates showed that the piezophilic bacteria are not widely distributed in terms of taxonomic positions, and all were assigned to particular branches of the Proteobacteria gamma-subgroup. A pressure-regulated operon from piezophilic bacteria of the genus *Shewanella*, *S. benthica* and *S. violacea*, was cloned and sequenced, and downstream of this operon another pressure regulated operon, *cydD-C*, was found. The *cydD* gene was found to be essential for the bacterial growth under high-pressure conditions, and the product of this gene was found to play a role in their respiratory system. Results obtained later indicated that the respiratory system in piezophilic bacteria may be important for survival in a high-pressure environment, and more studies focusing on other components of the respiratory chain have been conducted. These studies suggested that piezophilic bacteria are capable of changing their respiratory system in response to pressure conditions, and a proposed respiratory chain model has been suggested in this regard.

Introduction

The deep-sea is regarded as an extreme environment with conditions of high hydrostatic pressure [up to 110 megapascals (MPa)], and predominantly low temperature (1-2 °C), but with occasional regions of extremely high temperature (up to 375 °C) at sites of hydrothermal vents, darkness and low nutrient availability. It is accepted that deep-sea microbiology as a definable field did not exist before the middle of this century, and little attention was paid to this field except for the efforts of Certes and Portier (Jannasch and Taylor, 1984). Certes, during the Travaillier and Talisman Expeditions (1882-1883), examined sediment and water collected from depths to 5,000 m and found bacteria in almost every sample. He noted that bacteria survived at great pressure and might live in a state of suspended animation (Certes, 1884). In 1904, Portier used a sealed and autoclaved glass tube as a bacteriological sampling device and reported colony counts of bacteria from various depths and locations (Richard, 1907). In 1949, ZoBell and Johnson (1949) started work on the effect of hydrostatic pressure on microbial activities. The term "barophilic" was first used, defined as optimal growth at a pressure higher than 0.1 MPa or a requirement for increased pressure for growth. Recently, the term "piezophilic" was proposed to replace "barophilic" as the prefixes "baro" and "piezo", derived from Greek, mean "weight "and "pressure", respectively (Yayanos, 1995). Thus, the word piezophilic may be more suitable than barophilic to indicate bacteria that grow better at high pressure than at atmospheric pressure.

In this review, we have opted to use the term "piezophilic bacteria" which means high-pressure loving bacteria, and this review focuses on the taxonomy of deep-sea piezophiles and the features of their respiratory systems related to high pressure adaptation.

Taxonomic Positions of Deep-sea Piezophilic Bacteria

Bacteria living in the deep-sea display several unusual features that allow them to thrive in their extreme environment. The first pure culture of a piezophilic isolate, strain CNPT-3, was reported in 1979 (Yayanos *et al.*, 1979). This spirillum-like bacterium had a rapid doubling rate at 50 MPa, but no colonies were formed at atmospheric pressure even after incubation for several weeks. Numerous piezophilic and piezotolerant bacteria have since been isolated and characterized by the DEEPSTAR group at JAMSTEC, from deep-sea sediments at depths ranging from 2,500 m-11,000 m (Kato *et al.*, 1995a; 1996a; 1996b; 1998). Most of the isolated strains are not only piezophilic, but also psychrophilic, and they cannot be cultured at temperatures above 20 °C. The effects of pressure and temperature on cell growth, comparing the deep-sea piezophilic strains isolated by Yayanos and the DEEPSTAR group, are similar, in that all strains become more piezophilic at higher temperatures (Kato *et al.*, 1995a; Yayanos, 1986). These studies indicate that all piezophilic isolates are obligately piezophilic above the temperature at which growth occurs at atmospheric pressure. This means that the upper temperature limit for growth can be extended by high pressure. Likewise, piezophilic bacteria reproduce more rapidly at a lower temperature (such as 2 °C) when the pressure is less than that at its capture depth. It also appears to be true as a general rule that the pressure at which the rate of reproduction at 2 °C is maximal may reflect the true habitat depth of an isolate (Yayanos *et al.*, 1982).

Many deep-sea piezophilic bacteria have been shown to belong to the gamma-Proteobacteria through comparison of 5S and 16S rDNA sequences. As a result of a taxonomic study based on 5S rDNA sequences, Deming reported that the obligate piezophilic bacterium *Colwellia hadaliensis* belongs to the Proteobacteria gamma-subgroup (Deming *et al.*, 1988). DeLong *et al.* (1997) have also documented the existence of piezophilic and psychrophilic deep-sea bacteria that belong to this subgroup, as indicated by 16S rDNA sequences. The piezophilic bacteria reported by Liesack *et al.* (1991) are included in the same branch of the gamma-Proteobacteria as are the strains isolated by the DEEPSTAR group (Kato *et al.*, 1995a). These data suggest that most of the deep-sea high-pressure adapted piezophilic bacteria that can be readily cultured belong to the Proteobacteria gamma-subgroup, and these may not be widely distributed within the domain Bacteria.

DeLong *et al.* (1997) reported that eleven cultivated psychrophilic and piezophilic deep-sea bacteria are affiliated with one of five genera within the gamma-subgroup: *Shewanella*, *Photobacterium*, *Colwellia*, *Moritella*, and an unidentified genus. The only deep-sea piezophilic species of two of these genera were reported to be *S. benthica* in the genus *Shewanella* (Deming *et al.*, 1984; MacDonell and Colwell, 1985) and *C. hadaliensis* in the genus *Colwellia* (Deming *et al.*, 1988) prior to the reports from the DEEPSTAR group. We have identified four new piezophilic species within those genera based on the results of chromosomal DNA-DNA hybridization studies and several other taxonomic properties. Both previously described and new species of bacteria have been identified among the piezophilic bacterial isolates.

Photobacterium profundum, a new species, was identified through studies of the moderately piezophilic strains DSJ4 and SS9 (Nogi *et al.*, 1998a). *P. profundum* strain SS9 has been extensively studied with regard to the molecular mechanisms of pressure-regulation (Bartlett *et al.*, 1989; Bartlett *et al.*, 1996; Welch and Bartlett, 1998). *P. profundum* is the only species within the genus *Photobacterium* known to display piezophily, and the only one known to produce the long-chain polyunsaturated fatty acid (PUFA), eicosapentaenoic acid (EPA). No other species of *Photobacterium* produces EPA (Nogi *et al.*, 1998a).

The moderately piezophilic strain DSS12 isolated from the Ryukyu Trench at a depth of 5,110 m was identified as *Shewanella violacea* (Nogi *et al.*, 1998b), a novel piezophilic species belonging within the *Shewanella* piezophile branch (Li *et al.*, 1998). Other *Shewanella* piezophilic strains, PT-99 (DeLong and Yayanos, 1986), DB5501, DB6101, DB6705, DB6906 (Kato *et al.*, 1995a), DB172F, DB172R (Kato *et al.*, 1996b) and DB21MT-2 (Kato *et al.*, 1998) were all identified as members of the same species, *S. benthica* (Li *et al.*, 1998; Nogi *et al.*, 1998b; Nogi and Kato, 1999). The piezophilic and psychrophilic *Shewanella* strains, including *S. violacea* and *S. benthica*, also produce EPA (DeLong *et al.*, 1997; Kato *et al.*, 1998; Nogi *et al.*, 1998b), thus the occurrence of this PUFA is a property shared by many deep-sea bacteria. *Shewanella violacea* strain DSS12 has been well studied at JAMSTEC, particularly with respect to its molecular mechanisms of adaptation to high pressure. This strain is moderately piezophilic, with a fairly constant doubling time at pressures between 0.1 MPa and 70 MPa, whereas the doubling times of most of the piezophilic *S. benthica* strains change substantially with increasing pressure (Kato *et al.*, 1995a; 1996a; 1996b; 1998). Because there are few differences in the growth characteristics of strain DSS12 under different pressure conditions, this strain is a very convenient deep-sea bacterium for use in studies on the mechanisms of adaptation to high pressure environments. Studies using this strain include analyses of the pressure-regulation of gene expression (Kato et al., 1997a; 1997b) and of the role of d-type cytochromes in the growth of cells under high pressure (Kato et al., 1996c; Tamegai et al., 1998). The molecular mechanisms of gene expression have been analyzed focusing on a cloned pressure-regulated promoter and more detailed studies are in progress (Nakasone et al., 1998).

Strain DSK1, which is a moderately piezophilic bacterium isolated from the Japan Trench, was identified as Moritella japonica. This is the first piezophilic species identified in the genus Moritella (Nogi et al., 1998c). The type strain of the genus Moritella is M. marina, previously known as Vibrio marinus (Urakawa et al., 1998), one of the most common psychrophilic organisms isolated from marine environments (Colwell and Morita, 1964). M. marina is closely related to the genus Shewanella on the basis of 16S rDNA data, and is not a piezophilic bacterium (Nogi et al., 1998c). The extremely piezophilic bacterium strain DB21MT-5 isolated from the world deepest sea bottom, the Mariana Trench Challenger Deep, at a depth of 10,898 m (Kato et al., 1998), was identified as a Moritella species and designated M. yayanosii (Nogi and Kato, 1999). The optimal pressure for growth of M. yayanosii strain DB21MT-5 is 80 MPa, and this strain is not able to grow at pressures below 50 MPa, but it is able to grow well at higher pressures, even as high as 100 MPa (Kato et al., 1998). Production of the long chain PUFA, docosahexaenoic acid (DHA), is one of the characteristic properties of the genus Moritella (DeLong et al., 1997; Kato et al., 1998; Nogi et al., 1998c). DeLong and Yayanos (1985; 1986) reported that the fatty acid composition of the piezophilic strains changed as a function of pressure, and in general, greater amounts of PUFAs were synthesized at higher growth pressures. Approximately 70 % of the membrane lipids in M. yayanosii are PUFA, which is a finding consistent with its adaptation to very high pressures.

From the above studies, four more piezophilic species could be added to the list of identified deep-sea piezophiles shown in Table 1, and their phylogenetic positions, based on comparisons of 16 rDNA sequences, are also shown in Figure 1.

Pressure-regulation of Gene Expression and Protein Expression in Piezophilic Bacteria of the Genus Shewanella

A promoter, activated by growth at high pressure, was cloned from the piezophilic S. benthica strain DB6705 into E. coli (Kato et al., 1995b). Downstream from this promoter, two open reading frames (ORF1 and 2) were identified as one operon, designated as a pressure-regulated operon (Kato et al., 1996d). The highly conserved pressure-regulated operon from the moderately piezophilic S. violacea strain DSS12 was also cloned and sequenced (Kato et al., 1997a). Its sequence is almost identical to the operon from S. benthica strain DB6705. Downstream from this operon, another pressure-regulated operon was discovered whose first ORF was designated

Table 1. List of Identified Deep-sea Piezophilic Bacterial Species

Genus	Species	Strain	Properties	Reference
Colwellia	C. hadaliensis	BNL-1	Extremely piezophilic[1]	Deming et al., 1988
Photobacterium	P. profundum*	SS9, DSJ4	Moderately piezophilic[2]	Nogi et al., 1998a
Shewanella	S. benthica	PT-99, DB-series	Moderately, obligatorily and etremely piezophilic	Kato et al., 1998, Deming et al., 1984 Li et al., 1998 Macdonell & Colwell 1985 Nogi et al., 1998b
	S. violacea*	DSS12	Moderately piezophile	Nogi et al., 1998b
Moritella	M. japonica*	DSK1	Moderately piezophile	Nogi et al., 1998c
	M. yayanosii*	DB21MT-5	Extremely piezophile	Kato et al., 1998 Nogi and Kato, 1999

[1]Extremely piezophilic bacteria are defined as bacteria that are unable to grow at pressures below 50 MPa but able to grow well at 100 MPa.
[2]Moderately piezophilic bacteria are defined as bacteria displaying optimal growth at a pressure of less than 40 MPa, and which are able to grow well at atmospheric pressure.
*Novel deep-sea species reported by the DEEPSTAR group.

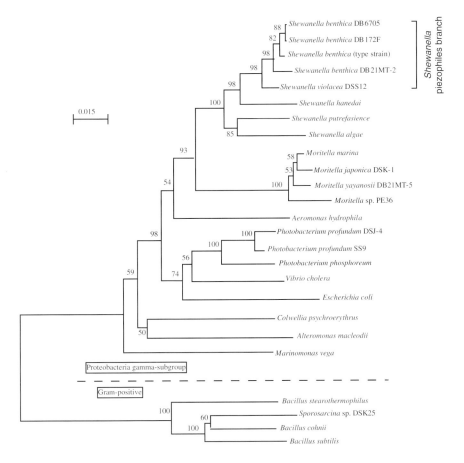

Figure 1. Phylogenetic Tree
Phylogenetic tree showing the relationships between isolated deep-sea barophilic bacteria within the gamma-subgroup of Proteobacteria and the genus Bacillus as determined by comparing 16S rDNA sequences using the neighbor-joining method. The scale represents the average number of nucleotide substitutions per site. Bootstrap values (%) are shown for frequencies above the threshold of 50%.

ORF3 and whose gene expression was also enhanced by high pressure (Kato *et al.*, 1997a). According to the results of transcriptional analyses, the pressure-regulated operons are expressed at elevated pressure, and at 70 MPa the transcripts are present in largest amount. Based upon the deduced amino acid sequence and the results of heterologous complementation studies, ORF3 appears to encode the CydD protein (Kato *et al.*, 1996c). In *E. coli*, CydD is required for the assembly of the cytochrome *bd* complex, one of the components of the aerobic respiratory chain (Poole *et al.*, 1994). *E. coli cydD* mutants display increased sensitivity to high pressure, but can be converted to cells which display wild type levels of high pressure sensitivity if the DSS12 ORF3 gene is introduced into the cells on a plasmid. In fact, the cytochrome *bd* protein complex of strain DSS12 was observed spectrophotometrically only in the case of cells grown under high pressure conditions (Tamegai *et al.*, 1998). It seems likely that the pressure regulation of this respiratory system in piezophilic bacteria plays a significant role in cell growth under high pressure conditions.

To study the pressure-regulated respiratory system in piezophilic *Shewanella* sp. in more detail, *c*-types cytochromes and a terminal quinol oxidase were purified and characterized from *S. benthica* strain DB172F (Qureshi *et al.*, 1998a; 1998b). Two kinds of cytochromes *c* from the membrane and the cytoplasm were purified, and named cytochrome *c*-551 and cytochrome *c*-552, respectively (Qureshi *et al.*, 1998a). Cytochrome *c*-551 was found to contain 44.2 nmol of heme *c* per mg of protein and cytochrome *c*-552 was found to contain 31.3 nmol of heme *c* per mg of protein. The CO difference spectrum of cytochrome *c*-551 showed a peak at 413.7 nm and troughs at 423.2, 522 and 552 nm which showed that this cytochrome combined with CO. Cytochrome *c*-551 was found to consist of two subunits with molecular masses of 29.1 kDa and 14.7 kDa, and each subunit contains a heme *c* molecule. Cytochrome *c*-552 was found to consist also of two subunits with molecular masses of 16.9 kDa and 14.7 kDa, and contains only one heme *c*. Cytochrome *c*-551 is constantly expressed during growth either 0.1 MPa or 60 MPa, whereas cytochrome *c*-552 is expressed only during growth at 0.1 MPa (Qureshi *et al.*, 1998a).

A novel membrane bound *ccb*-type quinol oxidase, from cells grown at 60 MPa, was purified to an electrophoretically homogenous state from the same strain (Qureshi *et al.*, 1998b). The purified enzyme complex was found to consist of four kinds of subunits with molecular masses of 98, 66, 18.5 and 15 kDa, and the complex contained 0.96 mol of protoheme and 1.95 mol of covalently bound heme *c* per mol of enzyme. Only the protoheme in the enzyme reacted with CO and CN^- and the catalytic activity of the enzyme was 50 % inhibited by 4 µM CN^-. The isoelectric point of the native enzyme complex was determined to be 5.0. This enzyme was found to be specifically induced only under elevated hydrostatic pressure conditions and it is expressed at high levels in cells grown at 60 MPa. In contrast, the membranes of cells grown at atmospheric pressure (0.1 MPa) exhibit high levels of cytochrome *c* oxidase activity and *N,N,N',N'*-tetramethyl-*p*-phenylenediamine ($TMPDH_2$)- oxidase activity (Qureshi *et al.*, 1998b).

These interesting observations suggest that the external growth pressure significantly alters the respiratory chain components, and furthermore, two

(A) AT 0.1 MPa PRESSURE

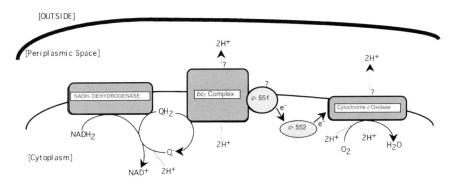

(B) AT 60 MPa PRESSURE

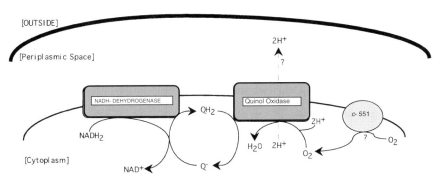

Figure 2. Electron Transport System in Piezophilic *Shewanella benthica*
Proposed electron transport system in piezophilic *Shewanella benthica* strain DB172F grown at 0.1 MPa (A) and 60 MPa (B). Abbreviations used are: Q⁻, quinone; QH₂, quinol; C-551, cytochrome *c*-551; C-552, cytochrome *c*-552.

kinds of respiratory chains regulated in response to pressure in the deep-sea *S. benthica* strain DB172F are present. A possible model of the respiratory system regulated by pressure is shown in Figure 2.

Description of the Respiratory System Model

Proposed Respiratory System Functioning at 0.1 MPa (Figure 2A)

In this model, three respiratory chain enzyme complexes are present, NADH-dehydrogenase (also called complex I), $bc1$- complex (also called complex III) and terminal cytochrome *c* oxidase (also called complex IV). Apparently, this is the typical respiratory chain found normally proposed in mitochondria and mesophilic bacteria. In this model, the flux of respiratory electrons occurs passing from complex I to complex IV. Pairs of high energy electrons derived from the metabolism of glucose are stored in the form of NADH and $FADH_2$. Complex I oxidizes $NADH_2$ to NAD and the two electrons released are transferred to quinone Q (oxidized form), which is then converted to the reduced state, quinol QH_2. Complex III then transfers the two electrons from quinol to membrane-bound cytochrom *c*-551. The soluble cytochrome *c*-552 then takes up the two electrons from *c*-551 and transfers them to the terminal oxidase. The terminal oxidase then reduces oxygen and pumps protons into the periplasmic space. The proton pump function of complex III is driven by the flow of electrons. Under normal physiological conditions, the donor of electrons to terminal oxidase is *c*-552, we suppose that *c*-551 might perform this function to involve with the complex III as an electron acceptor. However, further studies are needed in this regard.

Proposed Respiratory System Functioning at 60 MPa (Figure 2B)
At high pressure, the terminal oxidase enzyme in strain DB172F is quinol oxidase. The membrane-bound cytochrome *c*-551 is thought to have the ability to bind oxygen and supply it to the terminal quinol oxidase. This function is suggested by the finding that cytochrome *c*-551 is capable of binding with ligands such as carbon monoxide and cyanide. Hence, under elevated pressure conditions, the respiratory chain seems to be more compact and "short-cut" as compared to that under normal atmospheric pressure conditions. This may be intimately related to the property of bacterial piezophily. Being a new field of study, this area need will further exploration for the novel bioenergetic characteristics of deep-sea bacteria to be more fully understood.

Acknowledgements

We are very grateful to colleagues at JAMSTEC for providing deep-sea samples and for additional scientific cooperation, and to Dr. W. R. Bellamy for assistance in editing the manuscript.

References

Bartlett, D., Wright, M., Yayanos, A.A., and Silverman, M. 1989. Isolation of a gene regulated by hydrostatic pressure in a deep-sea bacterium. Nature. 342: 572-574.

Bartlett, D.H., Chi, E., and Welch, T. 1996. High pressure sensing and adaptation in the deep-sea bacterium *Photobacterium* species strain SS9. In: High Pressure Bioscience and Biotechnology. R. Hayashi and C. Balny, eds. Elsevier Science BV, The Netherlands. p. 29-36.

Certes, A. 1884. C. R. Acad. Sci. Paris. 98: 690-693.

Colwell, R.R., and Morita, R.Y. 1964. Reisolation and emendation of the description of *Vibrio marinus* (Russell) Ford. J. Bacteriol. 88: 831-837.

DeLong, E.F., and Yayanos, A.A. 1985. Adaptation of the membrane lipids of a deep-sea bacterium to changes in hydrostatic pressure. Science. 228: 1101-1103.

DeLong, E.F., and Yayanos, A.A. 1986. Biochemical function and ecological significance of novel bacterial lipids in deep-sea prokaryotes. Appl. Environ. Microbiol. 51: 730-737.

DeLong, E.F., Franks, D.G., and Yayanos, A.A. 1997. Evolutionary relationship of cultivated psychrophilic and barophilic deep-sea bacteria. Appl. Environ. Microbiol. 63: 2105-2108.

Deming, J.W., Hada, H., Colwell, R.R., Luehrsen, K.R., and Fox, G.E. 1984. The ribonucleotide sequence of 5S rRNA from two strains of deep-sea barophilic bacteria. J. Gen. Microbiol. 130: 1911-1920.

Deming, J.W., Somers, L.K., Straube, W.L., Swartz, D.G., and MacDonell, M.T. 1988. Isolation of an obligately barophilic bacterium and description of a new genus, *Colwellia* gen. nov. System. Appl. Microbiol. 10: 152-160.

Jannasch, H.W., and Taylor, C.D. 1984. Deep-sea microbiology. Annu. Rev. Microbiol. 38: 487-514.

Kato, C., Ikegami, A., Smorawinska, M., Usami, R., and Horikoshi, K. 1997a. Structure of genes in a pressure-regulated operon and adjacent regions from a barotolerant bacterium strain DSS12. J. Mar. Biotechnol. 5: 210-218.

Kato, C., Inoue, A., and Horikoshi, K. 1996a. Isolating and characterizing deep-sea marine microorganisms. Trends Biotechnol. 14: 6-12.

Kato, C., Li, L., Nakamura, Y., Nogi, Y., Tamaoka, J., and Horikoshi, K. 1998. Extremely barophilic bacteria isolated from the Mariana Trench, Challenger Deep, at a depth of 11,000 meters. Appl. Environ. Microbiol. 64: 1510-1513.

Kato, C., Li, L., Tamegai, H., Smorawinska, M., and Horikoshi, K. 1997b. A pressure-regulated gene cluster in deep-sea adapted bacteria with reference to its distribution. Recent Res. Devel. in Agricultural Biological Chem. 1: 25-32.

Kato, C., Masui, N., and Horikoshi, K. 1996b. Properties of obligately barophilic bacteria isolated from a sample of deep-sea sediment from the Izu-Bonin Trench. J. Mar. Biotechnol. 4: 96-99.

Kato, C., Sato, T., and Horikoshi, K. 1995a. Isolation and properties of barophilic and barotolerant bacteria from deep-sea mud samples. Biodiv. Conserv. 4: 1-9.

Kato, C., Smorawinska, M., Sato, T., and Horikoshi, K. 1995b. Cloning and expression in *Escherichia coli* of a pressure-regulated promoter region from a barophilic bacterium, strain DB6705. J. Mar. Biotechnol. 2: 125-129.

Kato, C., Smorawinska, M., Sato, T., and Horikoshi, K. 1996d. Analysis of a pressure-regulated operon from the barophilic bacterium strain DB6705. Biosci. Biotech. Biochem. 60: 166-168.

Kato, C., Tamegai, H., Ikegami, A., Usami, R., and Horikoshi, K. 1996c. Open reading frame 3 of the barotolerant bacterium strain DSS12 is complementary with *cydD* in *Escherichia coli*: *cydD* functions are required for cell stability at high pressure. J. Biochem. 120: 301-305.

Li, L., Kato, C., Nogi, Y., and Horikoshi, K. 1998. Distribution of the pressure-regulated operons in deep-sea bacteria. FEMS Microbiol. Lett. 159: 159-166.

Liesack, W., Weyland, H., and Stackebrandt, E. 1991. Potential risks of gene amplification by PCR as determined by 16S rDNA analysis of a mixed culture of strict barophilic bacteria. Microb. Ecol. 21: 191-198.

MacDonell, M.T., and Colwell, R.R. 1985. Phylogeny of the *Vibrionaceae*, and recommendation for two new genera, *Listonella* and *Shewanella*. System. Appl. Microbiol. 6: 171-182.

Nakasone, K., Ikegami, A., Kato, C., Usami, R., and Horikoshi, K. 1998. Mechanisms of gene expression controlled by pressure in deep-sea microorganisms. Extremophiles. 2: 149-154.

Nogi, Y., and Kato, C. 1999. Taxonomic studies of extremely barophilic bacteria isolated from the Mariana Trench, and *Moritella yayanosii* sp. nov., a new barophilic bacterial species. Extremophiles. 3: 71-77.

Nogi, Y., Kato, C., and Horikoshi, K. 1998c. *Moritella japonica* sp. nov., a novel barophilic bacterium isolated from a Japan Trench sediment. J. Gen. Appl. Microbiol. 44: 289-295.

Nogi, Y., Kato, C., and Horikoshi, K. 1998b. Taxonomic studies of deep-sea barophilic *Shewanella* species, and *Shewanella violacea* sp. nov., a new barophilic bacterial species. Arch. Microbiol. 170: 331-338.

Nogi, Y., Masui, N., and Kato, C. 1998a. *Photobacterium profundum* sp. nov., a new, moderately barophilic bacterial species isolated from a deep-sea sediment. Extremophiles. 2: 1-7.

Poole, R.K., Gibson, F., and Wu, G. 1994. The *cydD* gene product, componentof a heterodimeric ABC transporter, is required for assembly of periplasmic cytochrome *c* and of cytochrome *bd* in *Escherichia coli*. FEMS Microbiol. Lett. 117: 217-224.

Qureshi, M.H., Kato, C., and Horikoshi, K. 1998a. Purification of two pressure-regulated *c*-type cytochromes from a deep-sea barophilic bacterium, *Shewanella* sp. strain DB-172F. FEMS Microbiol. Lett. 161: 301-309.

Qureshi, M.H., Kato, C., and Horikoshi, K. 1998b. Purification of a ccb-type quinol oxidase specifically induced in a deep-sea barophilic bacterium, *Shewanella* sp. strain DB-172F. Extremophiles. 2: 93-99.

Richard, J. 1907. L'Oceanographie Paris: Vuibert and Nony.

Tamegai, H., Kato, C., and Horikoshi, K. 1998. Pressure-regulated respiratory system in barotolerant bacterium, *Shewanella* sp. strain DSS12. J. Biochem. Mol. Biol. Biophys. 1: 213-220.

Urakawa, H., Kita-Tsukamoto, K., Steven, S. E., Ohwada, K., and Colwell, R.R. 1998. A proposal to transfer *Vibrio marinus* (Russell 1891) to a new genus *Moritella* gen. nov. as *Moritella marina* comb. nov. FEMS Microbiol. Lett. 165: 373-378.

Welch, T.J., and Bartlett, D.H. 1998. Identification of a regulatory protein required for pressure-responsive gene expression in the deep-sea bacterium *Photobacterium* species strain SS9. Mol. Microbiol. 27: 977-985.

Yayanos, A.A. 1986. Evolutional and ecological implications of the properties of deep-sea barophilic bacteria. Proc. Natl. Acad. Sci. USA. 83: 9542-9546.

Yayanos, A.A. 1995. Microbiology to 10,500 meters in the deep sea. Annu. Rev. Microbiol. 49: 777-805.

Yayanos, A.A., Dietz, A. S., and Boxtel, R. V. 1979. Isolation of a deep-sea barophilic bacterium and some of its growth characteristics. Science. 205: 808-810.

Yayanos, A.A., Dietz, A. S., and Boxtel, R. V. 1982. Dependence of reproduction rate on pressure as a hallmark of deep-sea bacteria. Appl. Environ. Microbiol. 44: 1356-1361.

ZoBell, C.E., and Johnson, F.H. 1949. The influence of hydrostatic pressure on the growth and viability of terrestrial and marine bacteria. J. Bacteriol. 57: 179-189.

13

Microbial Adaptations to the Psychrosphere/Piezosphere

Douglas H. Bartlett

Center for Marine Biotechnology and Biomedicine
Scripps Institution of Oceanography
University of California, San Diego
La Jolla, CA 92093-0202, USA

Abstract

Low temperature and high pressure deep-sea environments occupy the largest fraction of the biosphere. Nevertheless, the molecular adaptations that enable life to exist under these conditions remain poorly understood. This article will provide an overview of the current picture on high pressure adaptation in cold oceanic environments, with an emphasis on genetic experiments performed on *Photobacterium profundum*. Thus far genes which have been found or implicated as important for pressure-sensing or pressure-adaptation include genes required for fatty acid unsaturation, the membrane protein genes *toxR* and *rseC* and the DNA recombination gene *recD*. Many deep-sea bacteria possess genes for the production of omega-3 polyunsaturated fatty acids. These could be of biotechnological significance since these fatty acids reduce the risk of cardiovascular disease and certain cancers and are useful as dietary supplements.

Introduction

Low temperature and high pressure environments (the psychrosphere and the piezosphere, respectfully) occupy the largest fractions of the known biosphere in terms of volume. Within these environments psychrophiles exist which can grow at temperatures as low as -12 °C (Russell, 1990), and piezophiles (also termed barophiles) exist which can grow at pressures as high as 130 megapascal [Yayanos, 1998; 1 megapascal (MPa) = 10 bar ≅ 9.9 atmospheres]. The lower temperature limit is apparently set by the freezing temperature of intracellular water, whereas the upper pressure limit for life is not known. Although Archaea, particularly those belonging to the nonthermophilic branch within the Crenarchaeota kingdom, are highly abundant in low temperature and high pressure environments (Fuhrman *et al.*, 1992; DeLong *et al.*, 1994), all of the cultured psychrotolerant/psychrophilic piezophilic prokaryotes fall within well known lineages of the Bacterial domain. Results to date indicate that the adaptations that enable these deep-sea microorganisms to grow under low temperature and high pressure conditions entail subtle variations of certain regulatory processes and aspects of cellular architecture present in mesophiles. Here I will briefly review genetic and biochemical studies of pressure adaptation by psychrotolerant/psychrophilic piezophilic prokaryotes with an emphasis on results obtained from my lab with *Photobacterium profundum*. Other reviews covering high pressure adaptation by low temperature-adapted deep-sea organisms include those written by Bartlett (1992), DeLong (1992), Prieur (1992), Somero (1990; 1991; 1992), Bartlett *et al.*, (1995), Yayanos (1995; 1998), Kato and Bartlett (1997), and within this miniseries symposium the article by Kato and Qureshi.

Basis of Pressure Effects

The thermodynamic consequences of changes in pressure are straightforward. Elevated pressure promotes decreased system volume changes associated with the equilibria and rates of biochemical processes. Thus, life in the deep sea must entail life at low volume change, or other biochemical adjustments which compensate for the consequences of elevated pressure. The relationship of pressure to equilibrium and rate processes can be expressed mathematically as follows:

$$1) \quad K_p = K_1 \exp(-P\Delta V/RT)$$

$$\text{and} \quad 2) \quad k_p = k_1 \exp(-P\Delta V^{\ddagger}/RT)$$

These two equations express the relationship of either the equilibrium constant (K) or rate constant (*k*) of a reaction to pressure, as determined by the size of the volume change that takes place during either the establishment of equilibrium (ΔV) or the formation of the activated complex ($\Delta V^{\ddagger}$). K_1 and k_1 are the equilibrium and rate constants, respectively, at atmospheric pressure; K_p and k_p are the constants at a higher pressure, R is the gas constant and T is absolute temperature. Because the equilibrium and rate constants are related to pressure in a logarithmic fashion, seemingly small changes in volume can lead to large changes in K or *k*. For example, a volume change of +50 cm^3mol^{-1} will lead to a 89% inhibition of a rate (or 89% change in K value) at 100 MPa pressure. A volume change of +100 cm^3mol^{-1} will lead to a 35% decrease in *k* or K at 10 MPa, and >99% decrease at 100 MPa. Volume changes accompanying many biological processes (both equilibria and reaction rates) are in the range of approximately 20-100 cm^3mol^{-1}. Therefore, pressure changes can lead to very substantial alteration (decrease or increase) of biological activities and structures in the absence of adaptation.

Because water itself is not highly compressible (about 4% per 100 MPa pressure), the effects of increased pressure on biochemical processes are not due to effects stemming from compression of the aqueous medium. However, water structure is very important in pressure sensitivity because changes in hydration of macromolecules and metabolites can lead to substantial alterations in system volume.

Changes in hydrostatic pressure perturb the volume of a system without changing its internal energy (as occurs with temperature), or its solvent composition (as occurs with pH and osmolarity changes). Thus, pressure is both a unique and a fundamental thermodynamic parameter.

Identification of Pressure Sensitive Processes in Mesophilic Microorganisms

Although the basis of pressure effects on many simple chemical processes have been described, pinpointing the physical nature of pressure effects on biological systems is extremely complicated. Changes in pressure, as with changes in other physical parameters, can affect a variety of cellular processes. Experimentally, cell division, flagellar function, and more critically, DNA replication, protein synthesis, enzyme function, and membrane structure have all been found to be sensitive to elevated pressure in shallow-water bacteria (reviewed in Bartlett, 1992). Many of these studies have been undertaken using the moderately piezotolerant *Escherichia coli*, where it has been found that pressures higher than 10 MPa inhibit motility, 20-50 MPa; cell division, 50 MPa; DNA replication, 58 MPa; protein synthesis, and 77 MPa; RNA synthesis.

When *Escherichia coli* was subjected to 55 MPa the number of polypeptides synthesized was greatly reduced, but many proteins exhibited elevated rates of synthesis relative to total protein synthesis (Welch *et al.*, 1993). As with other stress responses, this pressure-inducible protein (PIP) response was transient. The protein exhibiting the greatest pressure inducibility was a small basic protein not induced by any other known stress conditions. Among the PIPs identified in *E. coli* many present an interesting paradox. High pressure induces more heat shock proteins (eleven) than most other conditions outside of those which precisely mimic a heat shock response, while also inducing more cold shock proteins (four) than most conditions outside of those which precisely mimic a cold shock response. The high pressure induction of heat shock proteins has been observed in many prokaryotic and eukaryotic microorganisms as well as human cells in tissue culture (reviewed by Bartlett *et al.*, 1995).

Heat shock and cold shock proteins have inverse responses to a variety of conditions (VanBogelen and Neidhardt, 1990) so why should high pressure turn on both heat shock and cold shock proteins? Both cold temperature and high pressure inhibit an early step of protein synthesis. In addition, these conditions also both decrease membrane fluidity, increase nucleic acid secondary structure and either extremes of temperature as well as high pressure can reduce the stability of multimeric proteins (reviewed in Bartlett *et al.*, 1995) . The high pressure induction of heat shock and cold shock proteins helps to underscore the unique and pleiotropic consequences of shifts in pressure on cell functions. The relationship between high pressure and high temperature stress is further extended by the observation that high temperature preincubation of the yeast *Saccharomyces cerevisiae* confers significant protection against lethal levels of hydrostatic pressure (Komatsu *et al.*, 1991). How Hsps and other heat-inducible factors protect cells from high pressure stress is as yet unknown but could involve protecting membrane or protein stability.

Fatty Acid Regulation

Among the evolutionarily adaptive and transiently acclimatory changes thought to be important for growth at either low temperature or high pressure, membrane fatty acid alterations have received the most attention (Russell and Hamamoto, 1998; Yayanos, 1998). Both decreased temperature and increased pressure cause fatty acyl chains to pack together more tightly and influence gel state (L_β) to liquid crystalline (L_α) state transitions of membranes (reviewed by Bartlett and Bidle, 1999; Suutari and Laakso, 1994). For many phospholipids and natural membranes the transition slope in a temperature, pressure-phase diagram for the L_β - L_α main transition is approximately 20 °C/100 MPa at pressures lower than 100 MPa. Based on this relationship the combined effect of pressure and temperature on the phase state of a membrane from a deep-sea bacterium existing at 100 MPa and 2 °C is equivalent to an identical membrane at atmospheric pressure and -18 °C (lower than the low temperature limit for life). Decreased temperature or increased pressure have been found to result in increased unsaturation, chain length and the ratio of *anteiso* to *iso* branching, as well as decreased chain lengths. These changes may represent manifestations of "homeoviscous adaptation", the theory espoused by Sinensky (1974) whereby organisms adjust their membrane composition in response to changes in temperature in order to maintain a nearly constant membrane microviscosity. It has also been suggested that "homeophasic adaptation", maintaining a certain percentage of membrane phospholipids within a liquid crystalline state, is more physiologically relevant than the actual level of membrane fluidity (McElhaney, 1982).

    Although once thought to be virtually nonexistent in bacteria, omega-3 polyunsaturated fatty acids (PUFAs) are also frequently present in the lipids of microbes from low temperature and high pressure environments. There is an increased proportion of docosahexaenoic acid (DHA; 22:6) and/or eicosapentaenoic acid (EPA; 20:5) found in microbes isolated from deep-sea versus shallow water animals (Yano *et al.*, 1997). Higher percentages of EPA containing bacteria have also been noted in Antarctic versus temperate bacteria (Nichols *et al.*, 1993).

Photobacterium profundum as a Model System for Studying Pressure Adaptation

The only deep-sea isolate to yet be employed for genetic studies of pressure adaptation is *Photobacterium profundum* strain SS9. Members of this species have been isolated from amphipods in the Sulu Sea recovered from 2551 m

Table 1. Genes Cloned from Deep-Sea Piezotolerant/Piezophilic Bacteria

Gene	Protein/function	Source organism(s)	Isolation Depth (m)	Reference
asd	aspartate ß-D-semialdehyde dehydrogenase	Shewanella DSS12 and DB6705	5110 and 6356	(Kato, et al., 1997)
cydD, cydC	cytochrome bd biosynthesis	Assorted Shewanellas	up to 11000	(Kato, et al., 1995; Kato, et al., 1996; Kato, et al., 1997; Kato et al., 1997; Li et al., 1998)
mdh	malate dehydrogenase	P. profundum SS9	2551	(Welch and Bartlett, 1997)
ompH, ompL	probable porins	P. profundum SS9	2551	(Bartlett et al., 1993; Welch and Bartlett, 1996)
recD	Exonuclease V subunit	P. profundum SS9	2551	(Bidle and Bartlett, 1999)
rpsD, rpoA, rplQ	RNA polymerase a subunit and ribosomal proteins	Shewanella DB6705		(Nakasone et al., 1996)
rpoE operon	RNA polymerase s subunit and regulatory proteins	P. profundum SS9	2551	(Chi and Bartlett, 1995)
ssb	Single-stranded-DNA-binding protein	Various Shewanella strains	up to 8600	(Chilukuri and Bartlett, 1997)
toxRS operon	pressure sensors	P. profundum SS9	2551	(Welch and Bartlett, 1998)
ORF3/4 partial	EPA fatty acid synthase subunit	P. profundum SS9	2551	(Allen et al., 1999)

and from sediment obtained from 5110 m in the Ryukyu Trench (DeLong, 1986; Nogi *et al.*, 1998). SS9 has a relatively small genome size of 2 Mbp (Smorawinska and Kato; unpublished results), a temperature optimum of 15 °C and a pressure optimum of 20 - 30 MPa (DeLong, 1986). Why should this moderately piezophilic psychrotolerant species represent a useful model system for studies of life at high pressure and low temperature? It has a relatively rapid doubling time of 2 hours under optimal growth conditions and grows well over a wide temperature range (the 4 °C rate is nearly identical to the 15 °C rate) and pressure span (growth at 1 and 30 MPa are nearly identical and the upper pressure limit is 90 MPa). Furthermore, it is possible to introduce a variety of broad host range plasmids into SS9 by conjugal transfer, and to perform transposon mutagenesis (Tn5 and mini-Mu), interposon mutagenesis and allelic exchange (Chi and Bartlett, 1993; 1995; Bartlett and Chi, 1994; Welch and Bartlett, 1996; 1998). Finally, SS9 responds to pressure shifts by modulating the abundance of a number of fatty acids and proteins as described below. As can be seen in Table 1, many of the genes obtained thus far from piezophilic bacteria have been isolated from strain SS9.

Fatty Acid Regulation in *P. profundum*

As with many other deep-sea bacteria, strain SS9 enhances the proportion of both monounsaturated (MUFAs) and polyunsaturated fatty acids (PUFAs) when grown at decreased temperature or elevated pressure. In order to assess the importance of unsaturated fatty acids (UFAs) to microbial growth at low temperature and high pressure, the effect of treatments and mutations altering UFA levels in strain SS9 was investigated (Allen *et al.*, 1999). Treatment of cells with the ß-ketoacyl ACP synthase inhibitor cerulenin inhibited MUFA but not saturated fatty acid or PUFA synthesis, and led to decreased growth rate and yield at low temperature and high pressure in the absence of exogenous MUFA. The inspection of UFA auxotrophic mutants also indicated a role for MUFAs (particularly cis-vaccenic acid, 18:1) in growth of SS9 at low temperature and elevated pressure. One of these mutants, strain EA3, was deficient in the production of MUFAs and was sensitive to both low temperature and high pressure in the absence of exogenous 18:1 fatty acid. Another mutant, strain EA2, produced little MUFAs but elevated levels of the PUFA species eicosapentaenoic acid (EPA; 20:5n-3). This mutant grew slowly, but was not low temperature-sensitive or high pressure-sensitive. These results have provided the first evidence that UFAs are required for low temperature and high pressure adaptation, and have also suggested that overproduction of PUFAs might at least partially compensate for reduced MUFA levels. The fact that none of the MUFA mutants displayed reduced EPA levels further indicated that the production of these two types of UFAs are controlled by separate biosynthetic pathways.

The progress which has been made in recent years towards understanding the genetic basis of EPA synthesis has been used for directly testing the role of EPA in low temperature and high pressure growth of strain SS9. Much of the driving force for this research has been that omega-3 fatty acids such as EPA and DHA are considered important chemicals for

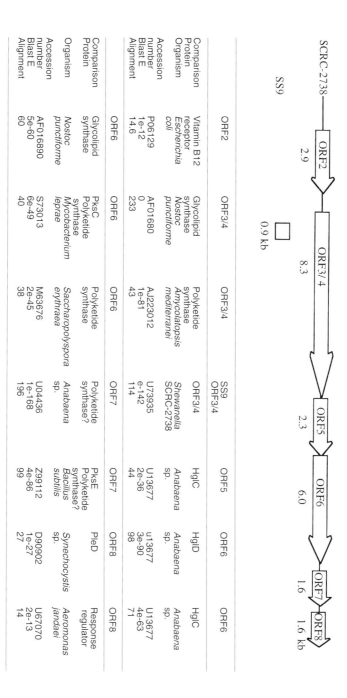

SCRC-2738 — ORF2 (2.9) — ORF3/4 (8.3) — ORF5 (2.3) — ORF6 (6.0) — ORF7 (1.6) — ORF8 (1.6 kb)

SS9 — 0.9 kb

	ORF2	ORF3/4	ORF3/4	SS9 ORF3/4	ORF5	ORF6
Comparison Protein	Vitamin B12 receptor	Glycolipid synthase	Polyketide synthase		HglC	HglC
Organism	Escherichia coli	Nostoc punctiforme	Amycolatopsis mediterranei	Shewanella SCRC-2738	Anabaena sp.	Anabaena sp.
Accession number	P06129	AF01680	AJ223012	U73935	U13677	U13677
Blast E	1e-12	0	1e-81	e-142	2e-36	3e-90
Alignment	14.6	233	43	114	44	71

	ORF6	ORF6	ORF7	ORF7	ORF8	ORF8
Comparison Protein	Glycolipid synthase	PksC Polyketide synthase	Polyketide synthase	Polyketide synthase?	PksE Polyketide synthase?	Response regulator
Organism	Nostoc punctiforme	Mycobacterium leprae	Saccharopolyspora erythraea	Anabaena sp.	Bacillus subtilis	Aeromonas jandaei
Accession number	AF016890	S73013	M63676	U04436	Z99112	U67070
Blast E	5e-60	6e-49	2e-45	1e-168	4e-86	2e-13
Alignment	60	40	38	196	99	14

Figure 1. The *Shewanella* EPA Biosynthetic Genes
Similarity of deduced protein sequences from *Shewanella* SCRC-2738 ORFs and one partial SS9 ORF involved in EPA biosynthesis with additional deduced protein sequences available in GenBank. The position, orientation and size of the ORFs are indicated below. ORFs 2-8 are of *Shewanella* SCRC-2738 (indicated as ORFS 6, 10, 12, 14, 16, 18 in Genbank U73935; Yazawa, 1996). The significance of the similarities are presented as gapped Blast E values (Altschul et al., 1997) and alignment scores derived from optimized values of RDF2 (Lipman and Pearson, 1985).

biotechnology because of their activities in reducing the risk of cardiovascular disease and certain cancers, their use as dietary supplements in mariculture and poultry farming, and their application as substrates for the synthesis of certain hormone-like molecules (Simopoulos, 1991). The molecular biology of PUFA production began when a 38 kbp genomic DNA fragment was isolated from a *Shewanella putrefaciens* species (neither psychrophilic or piezophilic) which conferred EPA production ability to both *E. coli* and *Synechococcus* (Yazawa, 1996; Takeyama *et al.*, 1997). The nucleotide sequence of this DNA was determined and based upon deletion analysis and deduced amino acid homology to other known enzymes involved in fatty acid synthesis, heterocyst glycolipid synthesis and polyketide antibiotic synthesis, six open reading frames were identified as important to EPA production (Yazawa, 1996). This genetic structure is very similar among diverse EPA and DHA producing bacteria (D. Facciotti, Calgene Inc.; personal communication). Because PUFA producing bacteria are present in many separate lineages, it is possible that the genes for PUFA biosynthesis have been acquired in many instances via horizontal gene transfer.   Figure 1 shows a schematic of the *S. putrefaciens* SCRC-2738 genes involved in EPA biosynthesis and the relationship of their gene products to other bacterial proteins.

Based on the sequence of the SCRC-2738 EPA genes it was possible to employ the polymerase chain reaction to amplify a portion of the SS9 EPA biosynthesis gene cluster and to use this DNA fragment to construct an EPA mutant using reverse genetics (Allen *et al.*, 1999). Much to our surprise this mutant displayed no growth defects whatsoever, including under conditions of low temperature and elevated pressure. Two explanations for this lack of altered phenotype could be that the increased MUFA content of the EPA mutant is capable of compensating for the absence of EPA or that EPA is only required under certain physiological conditions not evaluated in our work.

Another possibility is that EPA is not required for psychrotolerant or piezotolerant bacterial growth, but as a nutritional source used by higher organisms with which EPA producing microooorganisms have established symbiotic associations. Many of the PUFA producing microorganisms that have been discovered have been isolated from vertebrate or invertebrate sources (Yayanos *et al.*, 1979; Deming *et al.*, 1984; DeLong, 1986; DeLong and Yayanos, 1986; Yazawa, 1996; Yano *et al.*, 1997). Clearly there is a need to better define the roles and regulation of both MUFAs and PUFAs in deep-sea bacteria.

Pressure Regulation of Outer Membrane Proteins in *P. profundum*

Another pressure response of strain SS9 is to modulate outer membrane protein (OMP) abundance. When cells are shifted from atmospheric pressure to higher pressures the amount of the outer membrane porin-like protein OmpL is lessened and production of the porin-like protein OmpH is increased (Bartlett *et al.*, 1989; Chi and Bartlett, 1993; Welch and Bartlett, 1996; 1998). Although *ompL* mutants are not impaired in growth at 1 atm and *ompH* mutants are not altered in growth at elevated pressure, physiological experiments with various mutants suggest that OmpH provides a larger channel than OmpL (Bartlett and Chi, 1994), a property which could be important in the deep sea where nutrients are frequently limiting. The *ompL* and *ompH* genes are transcriptionally regulated by the inner membrane-localized ToxR and ToxS proteins (Welch and Bartlett, 1998). These proteins were first discovered in *Vibrio cholerae* where they act both as environmental sensors and regulators of gene expression in response to changes in temperature, osmolarity, pH and the levels of certain extracellular amino acids (Miller and Mekalanos, 1988; Gardel and Mekalanos, 1994). ToxR is an oligomeric protein whose cytoplasmic domain binds to genes under its control (Dziejman and Mekalanos, 1994; Otteman and Mekalanos, 1995). ToxS modulates ToxR activity (DiRita and Mekalanos, 1991).

SS9 ToxR/S pressure sensing appears to depend on the physical state of the inner membrane. Treatment of SS9 with local anesthetics at concentrations which are known to increase membrane fluidity results in a low-pressure ToxR/S signaling phenotype (high OmpL, low OmpH abundance) even when the cells are grown at high pressure (Welch and Bartlett, 1998). We hypothesize that the low pressure mode of ToxR/S regulation represents ToxR/S in its active state, where it is capable of binding to its cognate regulatory sequences, resulting in either repression or activation of gene expression.

High pressure decreases the abundance as well as specific activity of ToxR/S. ToxR (and presumably ToxS) levels decrease with increasing pressure, but even if ToxR/S levels are artificially increased at high pressure by cloning the genes on a multiple copy plasmid, elevated pressure still produces a wild type OMP pattern. In contrast, at increased temperatures ToxR abundance drops, but OmpL/H levels do not substantially vary, presumably because ToxR specific activity increases with temperature. Thus, the SS9 ToxR/S system appears well tuned to function as a barometer (or piezometer) rather than a thermometer. In contrast, ToxR/S dependent pressure regulation has not been observed in any of the several mesophilic *Vibrio* spp examined (our unpublished results).

Genes Required for Piezoadaptation in *P. profundum*

In addition to the *toxRS* operon two addition loci have been uncovered whose products influence OMP abundance. These remaining loci are particularly interesting because mutations in them render the cells extremely sensitive to elevated pressure.

The first locus which was found to be required for both psychro- and piezo-adaptation in strain SS9 was identified among SS9 transposon mutants which possessed reduced OmpH levels. Transposon cloning and DNA sequence analysis indicated that insertions in the *rseB* gene of the *rpoE* operon were responsible for these pressure-sensitive mutants (Chi and Bartlett, 1995). This operon has been investigated in several gram-negative bacteria, most notably *E. coli* and *Pseudomonas aeruginosa* (Deretic *et al.*, 1994; Lonetto, *et al.*, 1994; Raina *et al.*, 1995; Rouvière *et al.*, 1995; Ohman *et al.*, 1996; De Las Peñas *et al.*, 1997; Missiakas *et al.*, 1997). In these mesophiles the first gene in the operon, designated *rpoE* in the case of *E.*

coli, encodes an alternative sigma factor which activates the expression of a number of genes in response to damaging conditions in the extracytoplasmic compartment. The genes downstream from *rpoE* are designated *rseA*, *rseB* and *rseC* in *E. coli*. The *rseA* and *rseB* gene products regulate the activity of RpoE in response to the prevailing conditions within the membranes and periplasmic space. The last gene in the operon, *rseC*, is poorly understood. Mutations in *rseC* do not result in a dramatic change in RpoE activity in any of the mesophiles examined and *rseC* homologs have been found in several bacteria unlinked to *rpoE*. It is therefore believed that RseC functions independently of RpoE. Based on the phenotypes of *rseC* mutants in *Rhodobacter capsulatus* and *S. typhimurium* it has been suggested that RseC proteins function to promote the assembly of certain membrane-associated protein complexes (Beck *et al.*, 1997).

The SS9 *rseB* insertion mutants were altered in the abundances of numerous outer membrane proteins in addition to OmpH. The basis for this phenotype was not determined, but could have resulted from excessively high RpoE acitivity resulting from the inactivation of the *rseB* gene. *E. coli* RpoE controls the expression of the periplasmic protease DegP (Raina *et al.*, 1995; Rouvière *et al.*, 1995), whose activity could control the turnover of many outer membrane proteins.

High pressure revertants of the SS9 *rseB* mutants were readily obtained. These strains were always concomitantly restored for low temperature growth ability. Most of these revertants possessed DNA rearrangements at the site of the transposon insertion, further demonstrating the importance of the *rpoE* locus to high pressure and low temperature growth. Complementation analyses indicated that *rseC* is required for piezo- and psychro-adaptation. *rseC* mutants in *E. coli* and *P. aeruginosa* are neither low temperature-sensitive nor increased in piezosensitivity. Thus, while the SS9 *rpoE* operon is closely related to its homolog in mesophiles and therefore must function using essentially the same mechanisms present in other bacteria, the functioning of SS9 RseC seems particularly attuned to low temperature and high pressure conditions. Clarification of these distinctions could provide new insight into the function of the *rpoE* operon gene products and their physiological significance in bacteria adapted to different temperature (and pressure) regimes. As with MUFAs and ToxR/S, RseC represents another membrane-localized component needed for pressure-adaptation or pressure-signal transduction.

The second gene discovered to be required for piezo-adaptation of SS9 was isolated by complementation of a pressure-sensitive mutant, EC1002, impaired in the expression of a *ompH::lacZ* transcriptional fusion (Chi and Bartlett, 1993; Bidle and Bartlett, 1999). At high pressure this strain develops into enlarged spherical cells which divide poorly . The gene responsible for complementation of high pressure growth and normal cell morphology was found to encode a protein, RecD, which is involved in homologous DNA recombination. Further study of mutant EC1002 confirmed that it possessed phenotypes consistent with those of a *recD* mutant, i.e. a deficency in plasmid stability, and that this strain harbored a stop codon within the *recD* gene. Two additional *recD* mutants were constructed by gene disruption and were also found to possess a pressure-sensitive growth phenotypes.

The connection between DNA recombination, *ompH* expression and growth and normal cell morphology at high pressure is still unknown. However, it may be noteworthy that mutants impaired in chromosome partitioning have previously been observed to possess defects in outer membrane proteins (Bahloul *et al.*, 1996). RecD, along with RecB and RecC are subunits of Exonuclease V, which plays a major role in the homologous recombination pathway of bacteria; reviewed in Myers and Stahl (1994). RecBCD has been proposed to function as a destructive nuclease of linear DNA until it encounters the octameric sequence Chi (χ). At this sequence, RecD is hypothesized to dissociate from the complex, while the RecBC enzyme is proposed to convert to a χ-independent recombinogenic helicase, to produce single-stranded DNA that serves as a substrate for RecA-catalyzed recombination (Dixon *et al.*, 1994).

Because *E. coli recD* mutants are hyper-recombinogenic (Biek and Cohen, 1986), we propose that excessive DNA recombination is at the heart of SS9 *recD* mutant pressure-sensitivity. If a *recD* mutant in SS9 exhibits an increased frequency of inter-chromosomal recombination and genome concatemerization, then elevated pressure could compromise cell growth by exacerbating this condition. This could be accomplished by further increasing the rate of chromosome multimerization, or by inhibiting the rate of resolution of recombined chromosomes. Chromosome multimers would not segregate properly and would thereby block subsequent steps leading to cell division. Thus, impaired chromosome segregation could explain the poor growth and enlarged swollen cells of SS9 *recD* mutants grown at high pressure. Figure 2 presents a schematic of this model.

Another intriguing aspect of SS9 *recD* is its effect on *E. coli* cell morphology. Among the earliest observations of the effects of moderate pressure on bacteria, is that stressful pressures induce the development of highly filamentous cells (ZoBell and Cobet, 1963; ZoBell, 1970). The

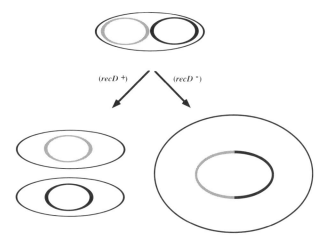

Figure 2. Schematic of the Possible Role of RecD in Preventing Interchromosomal Recombination in SS9 at Elevated Pressure

introduction of the SS9 *recD* gene into an *E. coli recD* mutant prevented cell filamentation at elevated pressures (Bidle and Bartlett, 1999). This effect was recessive to wild type *E. coli recD*.

Protein Adaptation

Studies of pressure effects on proteins from shallow and deep-sea animals indicate that deep-sea fishes and invertebrates possess proteins with structural and functional adaptations to high pressure (Somero, 1990; 1991; 1992). Unfortunately similar comparisons have not been undertaken for proteins from shallow and deep-sea bacteria, and no definitive data exists for the types of amino acid changes favored at high pressure. However, one comparison has been made of homologous proteins from related bacteria differing in pressure adaptation. This was performed for the oligomeric single-stranded DNA-binding protein (SSB) from four closely related marine shewanellas whose pressure optima ranged from atmospheric pressure to 69 MPa (Chilukuri and Bartlett, 1997). The *Shewanella* SSBs could be divided into conserved amino- and carboxy-terminal regions and a highly variable central region. Analysis of the amino acid composition of the *Shewanella* SSBs revealed several features that could correlate with pressure adaptation. A decrease in the proline and glycine content of the *Shewanella* SSBs, particularly in the central variable region of the protein, correlated strongly with the increased pressure adaptation of the source organism. The reduction in helix breaking residues (proline) and helix destabilizing residues (glycine) in the more piezophilic shewanellas indicates a decrease in flexibility of the protein, rendering the protein less compressible (Gross and Jaenicke, 1994) and possibly increasing its stability under pressure. Engineered mutants of staphylococcal nuclease which reduce chain flexibility increase protein stability at high pressure (Royer *et al.*, 1993).

In addition to intrinsic modifications of proteins from piezophilic bacteria, extrinsic factors could also be important in maintaining protein structure at high pressure. For example, small organic molecules act as thermostabilizers in some thermophiles (Hensel and König, 1988; Ciulla *et al.*, 1994; Martins and Santos, 1995). Recently, the concentration of certain osmolyte species present in strain SS9 have been found to increase with hydrostatic pressure (Martin, Bartlett and Roberts; manuscript submitted). Yancey and coworkers have likewise found that the organic osmolyte TMAO occurs at elevated levels in deep-sea versus shallow-water fishes (Gillett *et al.*, 1997). Furthermore, these investigators have found that TMAO reduces lactate dehydrogenase Km at high pressure. One possibility is that certain osmolytes which are excluded from the hydration layers of proteins could be used by deep-sea organisms to help counterbalance the protein hydrating influence of high pressure.

Concluding Remarks

There are opportunities for major discoveries to be made in deep-sea microbiology across a spectrum of levels of study extending from issues of biodiversity to the identification of genes required for piezoadaptation to membrane and protein biophysical studies at high pressure. With regard to the main focus of this review, the identification and elucidation of genes required for sensing and adapting to changes in hydrostatic pressure is still in its infancy. At this juncture it appears that genes which control membrane structure and DNA recombination are particularly important. But there are many possible pressure points in a bacterial cell, and many genetic modifications that must undoubtedly occur to enable life in the piezosphere.

Acknowledgements

D.H.B. gratefully acknowledges support from the National Science Foundation.

References

Allen, E.E, Facciotti, D., and Bartlett, D.H. 1999. Monounsaturated but not polyunsaturated fatty acids are required for growth at high pressure and low temperature in the deep-sea bacterium *Photobacterium profundum* strain SS9. Appl. Environ. Microbiol. 65: 1710-1720.

Altschul, S.F., Madden, T. L., Schaffer, A.A., Zhang, J., Zhang, Z., Miller, W., and Lipman, D.J. 1997. Gapped BLAST amd PSI-BLAST: a new generation of protein database search programs. Nucleic Acids Res. 25: 3389-3402.

Bahloul, A., Meury, J., Kern, R., Garwood, J., Guha, S., and Kohiyama, M. 1996. Co-ordination between membrane oriC sequestration factors and a chromosome partitioning protein, TolC (MukA). Mol. Microbiol. 22: 275-282.

Bartlett, D., Wright, M., Yayanos, A., and Silverman, M. 1989. Isolation of a gene regulated by hydrostatic pressure. Nature. 342: 572-574.

Bartlett, D.H. 1992. Microbial life at high pressures. Sci. Progress. 76: 479-496.

Bartlett, D.H., and Bidle, K.A. 1999. Membrane-based adaptations of deep-sea piezophiles. Enigmatic Microorganisms and Life in Extreme Environments. Kluwer Publishing Co., Dordrecht, The Netherlands. p. 503-512.

Bartlett, D.H., and Chi, E. 1994. Genetic characterization of *ompH* mutants in the deep-sea bacterium *Photobacterium* species strain SS9. Arch. Microbiol. 162: 323-328.

Bartlett, D.H., Chi, E., and Wright, M.E. 1993. Sequence of the *ompH* gene from the deep-sea bacterium *Photobacterium* SS9. Gene. 131: 125-128.

Bartlett, D.H., Kato, C., and Horokoshi, K. 1995. High pressure influences on gene and protein expression. Res. Microbiol. 697-706.

Beck, B.J., Connolly, L.E., De Las Peñas, A., and Downs, D.M. 1997. Evidence that *rseC*, a gene in the *rpoE* cluster, has a role in thiamine synthesis in *Salmonella typhimurium*. J. Bacteriol. 179: 6504-6508.

Bidle, K.A., and Bartlett, D.H. 1999. RecD function is required for high pressure growth in a deep-sea bacterium. J. Bacteriol. 181: 2330-2337.

Biek, D.P., and Cohen, S.N. 1986. Identification and characterization of *recD*, a gene affecting plasmid maintenance and recombination in *Escherichia coli*. J. Bacteriol. 167: 594-603.

Chi, E., and Bartlett, D.H. 1993. Use of a reporter gene to follow high pressure signal transduction in the deep-sea bacterium *Photobacterium* SS9. J. Bacteriol. 175: 7533-7540.

Chi, E., and Bartlett, D.H. 1995. An *rpoE*-like locus controls outer membrane protein synthesis and growth at cold temperatures and high pressures in the deep-sea bacterium *Photobacterium* SS9. Mol. Microbiol. 17: 713-726.

Chilukuri, L.N., and Bartlett, D.H. 1997. Isolation and characterization of the gene encoding single-stranded-DNA-binding protein (SSB) from four marine *Shewanella* strains that differ in their temperature and pressure optima for growth. Microbiol. 143: 1163-1174.

Ciulla, R.A., Burggraf, S., Stettr, K.O., and Roberts, M.F. 1994. Occurence and role of di-*myo*-inositol-1-1'-phosphate in *Methanococcus igneus*. Appl. Environ. Microbiol. 60: 3660-3664.

De Las Peñas, A., Connolly, L., and Gross, C.A. 1997. The σ^E-mediated response to extracytoplasmic stress in *Escherichia coli* is transduced by RseA and RseB, two negative regulators of σ^E. Mol. Microbiol. 24: 373-385.

DeLong, E.F. 1986. Adaptations of deep-sea bacteria to the abyssal environment. Ph.D. thesis. University of California, San Diego, La Jolla, Calif.

DeLong, E.F. 1992. High pressure habitats. In: Encyclopedia of Microbiology, vol. 2. J. Lederberg, ed. Academic Press Inc., San Diego. p. 405-417.

DeLong, E.F., Wu, K.Y., Prezelin, B.B., and Jovine, R.V.M. 1994. High abundance of archaea in Antarctic marine picoplankton. Nature. 371: 695-696.

DeLong, E.F., and Yayanos, A.A. 1986. Biochemical function and ecological significance of novel bacterial lipids in deep-sea prokaryotes. Appl. Environ. Microbiol. 51: 730-737.

Deming, J.W., Hada, H., Colwell, R.R., Luehrsen, K.R., and Fox, G.E. 1984. The ribonucleotide sequence of 5S rRNA from two strains of deep-sea barophilic bacteria. J. Gen. Microbiol. 130: 1911-1920.

Deretic, V., Schurr, M.J., Boucher, J.C., and Martin, D.W. 1994. Conversion of *Pseudomonas aeruginosa* to mucoidy in cystic fibrosis: environmental stress and regulation of bacterial virulence by alternative sigma factors. J. Bacteriol. 176: 2773-2780.

DiRita, V.J., and Mekalanos, J.J. 1991. Periplasmic interaction between two membrane regulatory proteins, ToxR and ToxS, results in signal transduction and transcriptional activation. Cell. 64: 29-37.

Dixon, D.A., Churchill, J.J., and Kowalczykowski, S.C. 1994. Reversible inactivation of the *Escherichia coli* RecBCD enzyme by the recombination hotspot χ *in vitro*: evidence for functional inactivation or loss of the RecD subunit. Proc. Natl. Acad. Sci. USA. 91: 2980-2984.

Dziejman, M., and Mekalanos, J.J. 1994. Analysis of membrane protein interaction: ToxR can dimerize the amino terminus of phage lambda repressor. Mol. Microbiol. 13: 485-494.

Fuhrman, J.A., McCallum, K., and Davis, A.A. 1992. Novel major archaebacterial group from marine plankton. Nature. 356: 148-149.

Gardel, C.L., and Mekalanos, J.J. 1994. Regulation of cholera toxin by temperature, pH, and osmolarity. Methods in Enzymol. 235: 517-527.

Gillett, M.B., Suko, J.R., Santoso, F.O., and Yancey, P.H. 1997. Elevated levels of trimethylamine oxide in muscles of deep-sea gadiform teleosts: A high-pressure adaptation? J. Exp. Zool. 279: 386-391.

Gross, M., and Jaenicke, R. 1994. Proteins under pressure - the influence of high hydrostatic pressure onstructure, function and assembly of protein complexes. Eur. J. Biochem. 221: 617-630.

Hensel, R., and König, H. 1988. Thermoadaptation of methanogenic bacteria by intracellular ion concentration. FEMS Microbiol. Lett. 49: 75-79.

Kato, C., and Bartlett, D.H. 1997. The molecular biology of barophilic bacteria. Extremophiles. 1: 111-116.

Kato, C., Ikegami, A., Smorawinska, M., Usami, R., and Horikoshi, K. 1997. Structure of genes in a pressure-regulated operon and adjacent regions from a barotolerant bacterium strain DSS12. J. Mar. Biotechnol. 5: 210-218.

Kato, C., Li, L., Tamaoka, J., and Horikoshi, K. 1997. Molecular analyses sediment of the 11000-m deep Mariana Trench. Extremophiles. 1: 117-123.

Kato, C., Li, L., Tamegai, T., Smorawinska, M., and Horikoshi, K. 1997. A pressure-regulated gene cluster in deep-sea adapted bacteria with reference to its distribution. Recent Res. Dev. Agricult., and Biol. Chem. 1: 25-32.

Kato, C., Smorawinska, M., Li, L., and Horikoshi, K. 1997. Comparison of the gene expression of aspartate ß-D-semialdehyde dehydrogenase at elevated pressure in deep-sea bacteria. J. Biochem. 121: 717-723.

Kato, C., Smorawinska, M., Sato, T., and Horikoshi, K. 1995. Cloning and expression in *Escherichia coli* of a pressure-regulated promoter region from a barophilic bacterium, strain DB6705. J. Mar. Biotech. 2: 125-129.

Kato, C., Smorawinska, M., Sato, T., and Horikoshi, K. 1996. Analysis of a pressure-regulated operon from the barophilic bacterium strain DB6705. Biosci. Biotech. Biochem. 60: 166-168.

Komatsu, Y., Obuchi, K., Iwahashi, H., Kaul, S.C., Ishimura, M., Fahy, G.M., and Rall, W.F. 1991. Deuterium oxide, dimethylsulfoxide and heat shock confer protection against hydrostatic pressure damage in yeast. Biochem. Biophys. Res. Commun. 174: 1141-1147.

Li, L., Kato, C., Nogi, Y., and Horikoshi, K. 1998. Distribution of the pressure-regulated operons in deep-sea bacteria. FEMS Microbiol. Lett. 159: 159-166.

Lipman, D.J., and Pearson, W.R. 1985. Rapid and sensitive protein similarity searches. Science. 227: 1435-1441.

Lonetto, M.A., Brown, K.L., Rudd, K.E., and Buttner, M.J. 1994. Analysis of the *Streptomyces coelicolor sigE* gene reveals the existence of a subfamily of eubacterial RNA polymerase s factors involved in the regulation of extracytoplasmic functions. Proc. Natl. Acad. Sci. USA. 91: 7573-7577.

Martins, L.O., and Santos, H. 1995. Accumulation of mannosylglycerate and di-myo-inositol phosphate by *Pyrococcus furiosus* in response to salinity and temperature. Appl. Environ. Microbiol. 61: 3299-3303.

McElhaney, R.N. 1982. Effects of membrane lipids on transport and enzymic activities. Current Topics in Membranes and Transport. Academic Press, New York. p. 317-380.

Miller, V.L., and Mekalanos, J.J. 1988. A novel suicide vector and its use in construction of insertion mutations: osmoregulation of outer membrane proteins and virulence determinants in *Vibrio cholerae* requires *toxR*. J. Bacteriol. 170: 2575-2583.

Missiakas, D., Mayer, M.P., Lemaire, M., Georgopoukos, C., and Raina, S. 1997. Modulation of the *Escherichia coli* σ^E(RpoE) heat-shock transcription factor activity by the RseA, RseB and RseC proteins. Mol. Microbiol. 24: 355-371.

Myers, R.S., and Stahl, F.S. 1994. Chi and the RecBCD enzyme of *Escherichia coli*. Annu. Rev. Genet. *28*: 49-70.

Nakasone, K., Kato, C., and Horikoshi, K. 1996. Molecular cloning of the gene encoding RNA polymerase alpha subunit from deep-sea barophilic bacterium. Biochim Biophys Acta 1308: 107-110.

Nichols, D.S., Nichols, P.D., and McMeekin, T.A. 1993. Polyunsaturated fatty acids in Antarctic bacteria. Antarctic Sci. 5: 149-160.

Nogi, Y., Masui, N., and Kato, C. 1998. *Photobacterium profundum* sp. nov., a new moderately barophilic bacterial species isolated from a deep-sea sediment. Extremophiles. 2: 1-7.

Ohman, D.E., Mathee, K., McPherson, C.J., DeVries, C.A., Ma, S., Wozniak, D.J., and Franklin, M.J. 1996. Regulation of the Alginate (*algD*) Operon in *Pseudomonas aeruginosa*. American Society for Microbiology, Washington, D.C.

Otteman, K.M., and Mekalanos, J.J. 1995. Analysis of *Vibrio cholerae* ToxR function by construction of novel fusion proteins. Mol. Microbiol. 15: 719-731.

Prieur, D. 1992. Physiology and Biotechnological Potential of Deep-sea Bacteria. Chapman Hall Inc., New York.

Raina, S., Missiakas, D., and Georgopoulos, C. 1995. The *rpoE* gene encoding the σ^E (σ^{24}) heat shock sigma factor of *Escherichia coli*. EMBO J. 14: 1043-1055.

Rouvière, P.E, De Las Peñas, A., Mecsas, J., Lu, C.Z., Rudd, K.E., and Gross, C.A. 1995. *rpoE*, the gene encoding the second heat-shock sigma factor, s^E, in *Escherichia coli*. EMBO J. 14: 1032-1042.

Royer, C.A., Hinck, A.P., Loh, S.N., Prehoda, K.E., Peng, X., Jonas, J., and Markley, J.L. 1993. Effects of amino acid substitutions on the pressure denaturation of Staphylococcal nuclease as monitored by fluorescence and nuclear magnetic resonance spectroscopy. Biochem. 32: 5222-5232.

Russell, N.J. 1990. Cold adaptation of microorganisms. Phil. Trans. R. Soc. London B 326: 595-611.

Russell, N.J., and Hamamoto, T. 1998. Psychrophiles. Extremophiles, Microbial Life in Extreme Environments. Wiley-Liss, New York. p. 25 - 45.

Simopoulos, A.P. 1991. Omega-3 fatty acids in health and disease and in growth and development. Am. J. Clin. Nutr. 54: 438-463.

Sinensky, M. 1974. Homeoviscous adaptation- a homeostatic process that regulates the viscosity of membrane lipids in *Escherichia coli*. Proc. Natl. Acad. Sci. USA. 71: 522-526.

Somero, G.N. 1990. Life at low volume change: hydrostatic pressure as a selective factor in the aquatic environment. Amer. Zool. 30: 123-135.

Somero, G.N. 1991. Hydrostatic Pressure and Adaptations to the Deep Sea. Wiley-Liss, Inc., New York.

Somero, G.N. 1992. Adaptations to high hydrostatic pressure. Ann. Rev. Physiol. 54: 557-577.

Suutari, M., and Laakso, S. 1994. Microbial fatty acids and thermal adaptation. Crit. Rev. Microbiol. 20: 285-328.

Takeyama, H., Takeda, D., Yazawa, K., Yamada, A., and Matsunaga, T. 1997. Expression of the eicosapentaenoic acid synthesis gene cluster from *Shewanella* sp. in a transgenic marine cyanobacterium. *Synechococcus* sp. Microbiol. 143: 2725-2731.

VanBogelen, R.A., and Neidhardt, F.C. 1990. Ribosomes as sensors of heat and cold shock in *Escherichia coli*. Proc. Natl. Acad. Sci. USA. 87: 5589-5593.

Welch, T.J., and Bartlett, D.H. 1996. Isolation and characterization of the structural gene for OmpL, a pressure-regulated porin-like protein from the deep-sea bacterium *Photobacterium* species strain SS9. J. Bacteriol. 178: 5027-5031.

Welch, T.J., and Bartlett, D.H. 1997. Cloning, sequencing and overexpression of the gene encoding malate dehydrogenase from the deep-sea bacterium *Photobacterium* species strain SS9. Biochim. Biophys. Acta 1350: 41-46.

Welch, T.J., and Bartlett, D.H. 1998. Identification of a regulatory protein required for pressure-responsive gene expression in the deep-sea bacterium *Photobacterium* species strain SS9. Mol. Microbiol. 27. In press.

Welch, T.J., Farewell, A, Neidhardt, F C and Bartlett, D H. 1993. Stress response in *Escherichia coli* induced by elevated hydrostatic pressure. J. Bacteriol. 175: 7170-7177.

Yano, Y., Nakayama, A., and Yoshida, K. 1997. Distribution of polyunsaturated fatty acids in bacteria present in intestines of deep-sea fish and shallow-sea poikilothermic animals. Appl. Environ. Microbiol. 63: 2572-2577.

Yayanos, A.A. 1995. Microbiology to 10,500 meters in the deep sea. Annu. Rev. Mirobiol. 49: 777-805.

Yayanos, A.A. 1998. Empirical and theoretical aspects of life at high pressure in the deep sea. Extremophiles. Microbial life in extreme environments. Wiley and Sons, Inc., New York. p. 47-92.

Yayanos, A.A., Dietz, A.S., and Van Boxtel, R. 1979. Isolation of a deep-sea barophilic bacterium and some of its growth characteristics. Science. 205: 808-810.

Yazawa, K. 1996. Production of Eicosapentaenoic acid from marine bacteria. Lipids 31: S-297-300.

ZoBell, C.E. 1970. Pressure Effects on Morphology and Life Processes of Bacteria. Academic Press, London and New York.

ZoBell, C.E and Cobet, A B. 1963. Filament formation by *Escherichia coli* at increased hydrostatic pressures. J. Bacteriol. 87: 710-719.

14

Adaptation of Proteins from Hyperthermophiles to Extreme Pressure and Temperature

Frank T. Robb[1], and Douglas S. Clark[2]

[1]Center of Marine Biotechnology, University of Maryland Biotechnology Institute, Baltimore, MD 21202, USA. [2]Department of Chemical Engineering, University of California, Berkeley, USA

Abstract

Further clarification of the adaptations permitting the persistence of life at temperatures above 100 °C depends in part on the analysis of adaptive mechanisms at the protein level. The hyperthermophiles include both Bacteria and Archaea, although the majority of isolates growing at or above 100 °C are Archaea. Newly described adaptive features of hyperthermophiles include proteins whose structural integrity persists at temperatures up to 200 °C, and under elevated hydrostatic pressure, which in some cases adds significant increments of stability.

Introduction

It is now very well established that microbial growth can occur at temperatures well above 100 °C. The recently described microorganism with the highest recorded growth temperature of 113.5 °C, is the chemolithoautotrophic archaeon *Pyrolobus fumarii* (Blochl *et al.*, 1997). Hyperthermophiles, which are defined as organisms with maximal growth temperatures above 90 °C (Adams, 1994), are widely distributed in hydrothermal environments in terrestrial as well marine and abyssal vent systems. The group includes 21 archaeal genera, and two bacterial genera with diverse growth physiology including both heterotrophs and chemoautotrophs (Stetter, 1996; Robb *et al.*, 1995). The archaeal hyperthermophiles have exceptionally high growth temperature limits, unique molecular characteristics, such as ether-linked lipids and eukaryotic transcription and translation factors (Robb *et al.* 1995), and unusual enzyme chemistry, such as the presence of multiple tungsten-containing enzymes (Kletzin and Adams, 1996). The unusual adaptive responses of the hyperthermophiles include heat shock proteins of the hsp60 family (Trent, 1996; Trent *et al.*, 1997; Kagawa *et al.*, 1995) which are inducible by supraoptimal temperatures, and reverse gyrase which installs positive supercoils in DNA (Guipaud *et al.*, 1997; Borges *et al.*, 1997).

The ongoing release of genomic sequence data from hyperthermophiles will continue to accelerate the discovery of thermostable proteins, however understanding the functional adaptations of these proteins requires the application of novel methods of analysis. For example, two key enzymes in glycolysis, phosphofructokinase and hexokinase, remained undiscovered in extracts of the hyperthermophile *Pyrococcus furiosus* because it was anticipated that they would require ATP. It transpires that these enzymes are active and highly expressed, but require ADP (Kengen *et al.*, 1995), presumably due to the higher stability of ADP compared with ATP. This review focuses on mechanisms of stabilization of enzymes with novel catalytic specificities, unusual thermal stability, and the ability to withstand, and in many cases to gain enhanced thermostability under elevated hydrostatic pressure. The latter is an unusual and intriguing feature of many proteins from thermophiles and recent work suggests that the structural stability of proteins in the 100-130 °C range is achieved by quite generic adaptive mechanisms. We will review the current field of protein stability at very high temperatures and discuss recent findings on the theoretical basis for pressure stabilization of proteins.

Mechanisms of Hyperstability

The known strategies for maintaining stability at high temperatures are summarized in Table 1. Five classes of adaptations are discussed below.

1. Amino Acid Composition
Although many formulae for compositional bias in hyperstable proteins have been suggested, there appear to be few answers to the quest for a thermoadapted amino acid composition. On the contrary, hyperstable proteins, like their counterparts in mesophiles, are composed of the same 20 amino acid set found in mesophiles. These amino acids undergo accelerated covalent modification at the high temperatures, pressures and extremes of pH that many thermophiles and hyperthermophiles can withstand (Hensel *et al.*, 1992). At temperatures beyond 100 °C, the stability of common amino acids declines in the following order:
(Val,Leu)>Ile>Tyr>Lys>His>Met>Thr>Ser> Trp>(Asp, Glu, Arg, Cys) (Jaenicke 1991; Jaenicke 1996a; Jaenicke and Bohm, 1998). Since many of the hyperthermophiles are relatively slow growing, the half lives of some of the more labile amino acids such as Cys, Arg, Gln, Asn are considerably shorter than the generation times of the organisms (Bernhardt *et al.*, 1984). Common types of chemical alterations include deamidation, ß-oxidation, Maillard reactions, hydrolysis, and disulfide interchange, etc (Jaenicke and Bohm, 1998). Many of the amino acids are more stable upon internalization in the hydrophobic core of the proteins than in free solution (Hensel *et al.*, 1992), however the frequencies of occurrence of Cys, Asn, Gln and Asp is

Table 1. Synopsis of Mechanisms of Extreme Thermostability in Proteins from Hyperthermophiles

Organism (Topt)	Thermostability feature	Protein	Reference
Pyrococcus furiosus (100 °C)	Mg^{++} requirement KCl (1M)	Acetyl-CoA synthetase (ADP-forming)	Glasemacher et al., 1997
P. furiosus (100 °C)	Salt bridges	Glutamate dehydrogenase	Klump et al., 1992; Vetriani et al., 1998
Sulfolobus acidocaldarius (85 °C)	Lysine monomethylation NOT salt bridges	Sac7d	McCrary et al., 1996
S. acidocaldarius (85 °C)	The data conforms to none of the sequence based conventions proposed by others	Adenylate kinase Pyrophosphatase Superoxide dismutase	Schafer et al., 1996
Methanothermus fervidus (65 °C)	Hydrophobic "proline N caps" Interhelical hydrogen bonds Short N- and C-termini	Histone rHMfB	Starich et al., 1996
Aquifex pyrophilus (98 °C)	Salt bridges Hydrophobic packing Polymeric tetramer	Superoxide dismutase (SOD)	Lim et al., 1997
Thermotoga maritima (77 °C)	Homodimer Replace loop with helix	Phosphoribosyl anthranilate isomerase	Hennig et al., 1997
T. maritima (77 °C)	Homo-dimer hydrophobic packing N- and C- termini immobilized Salt bridges	Phosphoribosyl anthranilate isomerase	Hennig et al., 1997

significantly lowered in the hyperstable variants of some common proteins. The genomic sequences that are determined or in progress will continue to make available a flood of deduced protein sequences, which will lead to additional insights in the area of compositional bias. Compatible solutes, some of which have recently been found to provide generic thermoprotection to proteins, may be acting by incrementally slowing amino acid modification reactions at high temperatures (Hensel *et al.*, 1992).

2. Hydrophobic Packing

An extensive, efficiently packed hydrophobic core is a common feature of stable globular proteins, although it appears to be more dominant in the lesser hyper-thermophiles than in those growing at or above 100 °C (Russell *et al.*, 1995; Macedo-Ribeiro, 1997; Pfeil *et al.*, 1997; Jaenicke, 1996b).

At present, the effect of temperature on hydrophobic interactions is still a subject of debate in the literature. The strength of the hydrophobic interaction is most commonly represented by ΔG°_{tr}, the Gibbs free energy of transfer of hydrocarbons from a pure hydrocarbon liquid to water (Schellman, 1997). An alternative is to define the strength of the hydrophobic interaction by $\Delta G^{\circ}_{tr}/T$, which is proportional to the natural log of the equilibrium constant (Schellman, 1997). The temperature dependence of the strength of the hydrophobic interaction depends on which definition is used. In particular, plots of ΔG°_{tr} vs. T for model hydrocarbons exhibit a maximum at around 140 °C (Privalov and Gill, 1988), whereas curves of $\Delta G^{\circ}_{tr}/T$ vs. T have a maximum at around 20 °C (Privalov and Gill, 1988). Thus, it is still unclear whether hydrophobic interactions in proteins become stronger or weaker at temperatures above room temperature, and how important they are to the stability of thermophilic proteins at high temperatures. Interestingly, some of the same constraints appear to operate in limiting the stability of proteins at low temperatures, between −10 and 20 °C (Privalov and Gill, 1988). There is also evidence that hyperstable glutamate dehydrogenase requires high temperatures for assembly, suggesting that specific hydrophobic states are critical to the molecular engagement of subunits, and that partly assembled multimeric proteins may be "frozen" in metastable intermediates. This system may provide insights into the nature of the folding and assembly pathways of hyperstable proteins (DiRuggiero and Robb, 1995; 1996).

3. Ionic Networks

Extensive networks of acidic and basic side chains on the surface of subunits and domains interact to form cooperatively bound assemblages (Russell and Taylor, 1995; Robb and Maeder, 1998; Rice *et al.*, 1996; Britton *et al.*, 1995). Ionic interactions by nature act over much longer ranges than hydrophobic interactions, and are relatively immune to alterations in the structure of water which is compromised at elevated temperatures. Consequently, widespread networks of ionic interactions are observed in the proteins of the more extreme hyperthermophiles compared with homologous proteins in the thermophiles (T_{opt} for growth between 50 and 75°C) or in mesophiles (Osterdorp *et al.*, 1996; Britton *et al.*, 1995). The ionic networks of *P. furiosus* glutamate dehydrogenase demonstrate extensive clustering of spatially alternating positive and negative charges, which have a component of hydrogen bonding. Recreating a network in the less stable glutamate dehydrogenase from *Thermococcus litoralis* resulted in elevated thermostability without any penalty in catalytic activity (Vetriani *et al.*, 1998).

4. Cooperative Association

In proteins that form oligomers or bind to larger substrates than themselves, dissociation is thought to precede the irreversible unraveling of monomers. The loss of integrity of the protein molecule is dependent on the unfolding of the monomeric protein into the denatured form. It follows that strong intermolecular associations can forestall this process. Consequently, many proteins that are monomeric in mesophiles are found to form oligomers in extreme thermophiles or hyperthermophiles. For example, the chorismate mutase from the hyperthermophile, *M. jannaschii*, appears to have developed a dimeric quaternary organization as an adaptation to stability (MacBeath *et al.*, 1998), and the compact beta-alpha TIM barrel of phosphoribosyl anthranilate isomerase, which is monomeric in enteric bacteria, is dimeric in *T. maritima* (Hennig *et al.*, 1997),

5. Pinning the Loose Ends

It is thought that many N- and C-termini are immobilized in hyperstable proteins, and are therefore not free to fray and initiate unraveling. In the case of phosphoribosyl anthranilate isomerase from *T.maritima* (Hennig *et al.*, 1997), not only are the termini buried in hydrophobic pockets, but a disordered loop found in the homologous protein from *E. coli* is replaced with an alpha helix, thereby inhibiting the nucleation of melting. The C-termini of the dimeric citrate synthases of hyperthermophiles has an intertwined structure which contributes stability by locking the dimer together as well as securing the carboxyl groups by ionic interaction, preventing disengagement of the ends (Russell *et al.*, 1997). The protein with the highest temperature stability on record, *P. furiosus* ferredoxin, which retains structure at temperatures up to 200 °C (Hiller *et al.*, 1997), features an ion-pair at the amino terminus (Cavagnero *et al.*, 1998). Loose ends must be handled somewhat differently in the multiple proteins from hyperthermophiles that have inteins. For example, the ribonucleotide reductase from *Pyrococcus furiosus* has been shown to have two inteins (Riera *et al.*, 1997), and the mechanisms of intein excision have been shown to proceed at optimal growth temperatures (Perler *et al.*, 1997), implying that there is an end-stabilizing mechanism for these protein splicing reactions.

Pressure Stabilization

Many hyperthermophiles isolated from deep sea hydrothermal vents are either indifferent to the effects of pressure on growth at high temperature, or else they are barophilic in terms of maximal growth rate and the upper temperature limits of growth (Miller *et al.*, 1988; Nelson *et al.*, 1992; Pledger *et al.*, 1994; Marteinsson *et al.*, 1997). Although many studies of pressure effects on enzyme stability have appeared since the early 1950's, in nearly all cases the experiments were performed with proteins from mesophilic sources at temperatures below 60 °C, such as lysozyme (Samarasighe *et al.*, 1992). Moreover, much of this early work focused on enzyme denaturation at very high pressures (> 300 MPa) (Jaenicke, 1991; Weber and Drickamer, 1983). More recent work, however, has shown that moderate pressures (100 MPa) below those normally needed for pressure-induced denaturation can in fact stabilize proteins against thermoinactivation (Hei and Clark, 1994; Michels and Clark, 1997; Michels *et al.*, 1996; Mozhaev *et al.*, 1996). Particularly large effects have been observed for thermophilic enzymes at very high temperatures. This behavior has important implications for the adaptation of thermophilic proteins *in extremis*, as discussed more fully below.

In the first report of thermophilic enzyme stabilization by pressure, Hei and Clark (1994) examined the effect of pressure on the thermal stability of four partially purified hydrogenases from methanogens of the genus *Methanococcus*. Only one of these organisms, *M. jannaschii*, was isolated from a deep-sea habitat: a deep-sea hydrothermal vent at a depth of 2500 m (Jones *et al.*, 1983). Notably, hydrogenases from the extreme thermophiles *M. jannaschii* and *M. igneus* were substantially stabilized by pressure whereas hydrogenases from *M. thermolithotrophicus* (a moderate thermophile) and *M. maripaludis* (a mesophile) were destabilized by pressure. These results were the first demonstration of pressure stabilization for thermophilic enzymes, and showed that the effect is not unique to enzymes isolated from high-pressure environments. The work of Summit *et al.* (1998) confirmed that the DNA polymerases from non-barophilic thermophiles could be stabilized by pressure. Furthermore, the hydrogenase studies, in combination with studies on the effects of pressure on several homologous glyceraldehyde-3-phosphate dehydrogenases from mesophilic and thermophilic sources and a rubredoxin from *P. furiosus*, implicated hydrophobic interactions as an important factor in the stabilization of thermophilic enzymes by pressure.

In follow-up work to the studies of Hei and Clark (1994), Michels and Clark (1997) isolated a proteolytic enzyme from *M. jannaschii* and found that the enzyme was both activated and stabilized by pressure. Even at moderate pressures, the thermal half-life of the enzyme was exceptionally high; for example, $t_{1/2}$ = 45 min at 116 °C, and 7 min at 125 °C (at ~10 atm pressure). Moreover, the thermostability of the *M. jannaschii* protease increased with pressure in a physiologically-relevant range. For example, by raising the pressure to 500 atm, the half-life of the enzyme was increased 2.7-fold at 125 °C. The pronounced effect of pressure on the *M. jannaschii* protease stability was unusual for proteolytic enzymes: when 500 atm was applied to trypsin, α-chymotrypsin, and subtilisin Carlsberg, half-lives at the enzyme's reported melting temperature increased by 40%, 30% and 30%, respectively, compared to 170% for *M. jannaschii* protease at 125 °C (Michels and Clark, 1997).

The barophilic behavior of the *M. jannaschii* protease (and of the corresponding hydrogenase; Miller *et al.*, 1989) is especially noteworthy in view of the unusual barophily exhibited by *M. jannaschii* in high-pressure growth experiments (Miller *et al.*, 1988). Whether such pressure-activation and stabilization of key enzymes is responsible for the barophilic growth of *M. jannaschii* remains to be seen; however, the similar effects of pressure on growth and enzyme activity suggests that these phenomena are interrelated and motivates further pressure studies of enzymes from *M. jannaschii* and other deep-sea thermophiles.

In more recent work, Sun *et al.* (1999) examined the thermostability and pressure-induced thermostabilization of glutamate dehydrogenase (GDH) from the hyperthermophilic archaeon *Pyrococcus furiosus* (*Pf*) at temperatures up to 109 °C (*Pf* GDH, a hexamer composed of six identical subunits, has a melting temperature for denaturation of 113 °C). The native GDH from *Pf* was sub-stantially stabilized by 500 atm, up to 18-fold at 109 °C. By comparison, a recombinant GDH mutant containing an extra tetrapeptide at the C-terminus was stabilized to an even greater degree, by 28-fold at 105 °C. Although the presence of the tetrapeptide destabilized the enzyme markedly, the destabilizing effect was largely reversed by pressure. The remarkable degree of pressure stabilization, especially of the recombinant GDH mutant, could not be attributed to hydrophobic interactions alone. However, further insights into the possible mechanism(s) of pressure stabilization were provided by inactivation experiments in the presence of glycerol. Specifically, stabilization was also achieved by adding glycerol, albeit to a lesser extent than effected by pressure, suggesting that compression and/or rigidification of the protein's structure played a role in pressure-induced thermostabilization.

Of the primary intramolecular interactions responsible for maintaining native protein structure, only hydrophobic interactions are expected to be stabilized by elevated pressure (although this conclusion is not universally accepted in the literature). This effect has been attributed to the relatively open structure of water solvating apolar surfaces exposed during protein unfolding (Hei and Clark, 1994; Michels *et al.*, 1996). In contrast, intraprotein electrostatic interactions ("salt bridges" or ion pairs) are generally believed to be strongly destabilized by increasing pressure (Michels *et al.*, 1996). Hydrogen bonds are slightly stabilized by pressure (Michels *et al.*, 1996), but intramolecular hydrogen bonds readily exchange with solvating water molecules upon protein unfolding. Hydrogen bonds are therefore expected to have little net influence on the stability of proteins under pressure.

So far it appears that pressure-stabilization can result from the unfavorable thermodynamics of hydrophobic solvation and/or stabilizing interactions in the protein induced by compression or rigidification of susceptible structures. Although these effects are not restricted to proteins from organisms of the deep sea or other high-pressure environments, pressure-induced activation and stabilization of proteins might be an underlying characteristic of microbial barophily. Moreover, in some cases of pronounced stabilization, structural interactions within thermophilic proteins seemingly at odds with pressure stabilization, *e.g.* the extensive ionic pairing in GDH from *P. furiosus*, are apparently outweighed by other factors that contribute to net stabilization (Clark *et al.*, 1996). Clearly, much more needs to be learned about the mechanisms of pressure stabilization at high temperatures, and the apparent synergy between pressure and thermostability in proteins from hyperthermophiles. This information will no doubt provide new insights into structure-stability relationships in general, if not the evolutionary pressures and pathways that lead to them.

Conclusions

The discovery of a hyperthermophile that grows at 113 °C has implications for adaptive features of proteins, and confirms that, *in vivo*, enzyme stability and activity may be observed at temperatures above 110 °C. Intrinsically stable proteins are typical of hyperthermophiles and the studies of chemical stability at temperatures higher than 110 °C will become familiar experimental procedures in future. The current focus on ion-pair network formation as a critical mechanism for stabilizing protein structures above 100 °C may explain many instances of stabilization by optimization of subunit interactions. The effects of pressure in stabilizing thermostable proteins are intriguing and a mechanistic explanation may be close at hand.

Acknowledgements

Research in the author's laboratories is supported by grants BES 9410687 and BES9604961 from the National Science Foundation. We thank M.M. Sun for assistance in preparing the manuscript.

References

Adams, M.W. 1994. Biochemical diversity among sulfur-dependent, hyperthermophilic microorganisms. FEMS Microbiol. Rev. 15: 261-277.

Adams, M.W., and Zhou, Z.. 1997. Site-directed mutations of the 4Fe-ferredoxin from the hyperthermophilic archaeon *Pyrococcus furiosus*: role of the cluster-coordinating aspartate in physiological electron transfer reactions. Biochem. 36: 10892-10900.

Bernhardt, R., Makower, A., Janig, G.R., and Ruckpaul, K. 1984. Selective chemical modification of a functionally linked lysine in cytochrome P-450 LM2. Biochim. Biophys. Acta. 785: 186-190.

Blochl, E., Rachel, R., Burggraf, S., Hafenbradl, D., Jannasch H.W., and Stetter, K.O. 1997. *Pyrolobus fumarii*, gen. and sp. nov., represents a novel group of archaea, extending the upper temperature limit of life to 113 °C. Extremophiles. 1: 114-121.

Borges, K.M., Bergerat, A., Bogert, A.M., DiRuggiero, J., Forterre, P., and Robb, F.T. 1997. Characterization of the reverse gyrase from the hyperthermophilic archaeon *Pyrococcus furiosus*. J. Bacteriol. 179: 1721-1726.

Britton, K.L., Baker, P.J., Borges, K.M., Engel, P.C., Pasquo, A., Rice, D.W., Robb, F.T., Scandurra, R., Stillman, T.J., and Yip, K.S. 1995. Insights into thermal stability from a comparison of the glutamate dehydrogenases from *Pyrococcus furiosus* and *Thermococcus litoralis*. Eur. J. Biochem. 229: 688-695.

Cavagnero, S., Debe, D.A., Zhou, Z..H., Adams, M.W., and Chan, S.I. 1998. Kinetic role of electrostatic interactions in the unfolding of hyperthermophilic and mesophilic rubredoxins. Biochem. 37: 3369-76

Clark, D.S., Sun, M., Giarto, L., Michels, P.C., Matschiner, A., and Robb, F.T. 1996. Activation and stabilization of thermophilic enzymes by pressure. In: High Pressure Bioscience and Biotechnology. R. Hayashi and C. Balny, eds. Elsevier Science Publishing Co., NY.

DiRuggiero, J., and Robb, F.T. 1995. Expression and *in vitro* assembly of recombinant glutamate dehydrogenase from the hyperthermophilic archaeon *Pyrococcus furiosus.* Appl. Environ. Microbiol. 61: 159-164.

DiRuggiero, J., and Robb, F.T. 1996. Enzymes of central nitrogen metabolism from hyperthermophiles: characterization, thermostability, and genetics. Adv. Protein Chem. 48: 311-339.

Glasemacher, J., Bock, A.K., Schmid, R., and Schonheit, P. 1997. Purification and properties of acetyl-CoA synthetase (ADP-forming), an archaeal enzyme of acetate formation and ATP synthesis, from the hyperthermophile *Pyrococcus furiosus.* Eur. J. Biochem. 244: 561-567

Guipaud, O., Marguet, E., Noll, K.M., de la Tour, C.B., and Forterre, P. 1997. Both DNA gyrase and reverse gyrase are present in the hyperthermophilic bacterium *Thermotoga maritima.* Proc. Natl. Acad. Sci. USA. 94: 10606-10611.

Hei, D.J., and Clark, D.S. 1994. Pressure stabilization of proteins from extreme thermophiles. Appl. Environ. Microbiol. 60: 932-939.

Hennig, M., Sterner, R., Kirschner, K., and Jansonius, J.N. 1997. Crystal structure at 2.0 A resolution of phosphoribosyl anthranilate isomerase from the hyperthermophile *Thermotoga maritima*: possible determinants of protein stability. Biochem. 36: 6009-6016.

Hensel, R., Jakob, I., Scheer, H., and Lottspeich, F. 1992. Proteins from hyperthermophilic archaea: stability towards covalent modification of the peptide chain. Biochem. Soc. Symp. 58: 127-33.

Hiller, R., Zhou, Z..H., Adams, M.W., and Englander, S.W. 1997. Stability and dynamics in a hyperthermophilic protein with melting temperature close to 200 °C. Proc. Natl. Acad. Sci. USA. 94: 11329-11332.

Holden, J.F., and Baross, J.A. 1995. Enhanced thermotolerance by hydrostatic pressure in the deep-sea hyperthermophile *Pyrococcus* strain ES4. FEMS Microbiol. Ecol. 18: 27-33.

Jaenicke, R. 1991. Protein stability and molecular adaption to extreme conditions. Eur. J. Biochem. 202: 715-728.

Jaenicke, R. 1996a. Structure and stability of hyperstable proteins: glycolytic enzymes from hyperthermophilic bacterium *Thermotoga maritima.* Adv. Protein. Chem. 48: 181-269.

Jaenicke, R. 1996b. Glyceraldehyde-3-phosphate dehydrogenase from *Thermotoga maritima*: strategies of protein stabilization. FEMS Microbiol. Rev. 18: 215-224.

Jaenicke, R., and Bohm, G. 1998. The stability of proteins in extreme environments. Curr. Opin. Struct. Biol. 8: 738-48.

Jones, W., Leigh, A., Mayer, F., Woese, C.R., and Wolfe, R. 1983. Arch. Microbiol. 136: 254-261.

Kagawa, H.K., Osipiuk, J., Maltsev, N., Overbeek, R., Quaite-Randall, E., Joachimiak, A., and Trent, J.D. 1995. The 60 kDa heat shock proteins in the hyperthermophilic archaeon *Sulfolobus shibatae.* J. Mol. Biol. 253: 712-725.

Kengen, S.W., Tuininga, J.E., de Bok, F.A., Stams, A.J., and de Vos, W.M. 1995. Purification and characterization of a novel ADP-dependent glucokinase from the hyperthermophilic archaeon *Pyrococcus furiosus.* J.

Biol. Chem. 270: 30453-30457.

Kletzin, A., and Adams, M.W. 1996. Tungsten in biological systems. FEMS Microbiol. Rev., 18: 5-63.

Klump, H., Di Ruggiero, J., Kessel, M., Park, J.B., Adams, M.W., and Robb, F.T. 1992. Glutamate dehydrogenase from the hyperthermophile *Pyrococcus furiosus*. Thermal denaturation and activation. J. Biol. Chem. 267: 22681-22685.

Lim, J.H., Yu, Y.G., Han, Y.S., Cho, S., Ahn, B.Y., Kim, S.H., and Cho, Y. 1997. The crystal structure of an Fe-superoxide dismutase from the hyperthermophile *Aquifex pyrophilus* at 1.9 A resolution: structural basis for thermostability. J. Mol. Biol. 270: 259-274.

MacBeath, G., Kast, P., and Hilvert, D. 1998. Redesigning enzyme topology by directed evolution. Science. 279: 1958-1961.

Macedo-Ribeiro, S., Darimont, B., and Sterner, R. 1997. Structural features correlated with the extreme thermostability of 1[4Fe-4S] ferredoxin from the hyperthermophilic bacterium *Thermotoga maritima*. J. Biol. Chem. 378: 331-336.

Marteinsson, M.T., Moulin, P., Birrien, J.-L., Gambacorta, A., Vernet, M., and Prieur, D. 1997. Physiological responses to stress conditions and barophilic behavior of the hyperthermophilic vent Archaeon *Pyrococcus abyssi*. Appl. Env. Micro. 63. 1230-1236.

McCrary, B.S., Edmondson, S.P., and Shriver, J.W. 1996. Hyperthermophile protein folding thermodynamics: differential scanning calorimetry and chemical denaturation of Sac7d. J. Mol. Biol. 264: 784-805.

Michels, P.C., and Clark, D.S. 1992. Pressure dependence of enzyme catalysis. In: Biocatalysis at Extreme Temperatures. M.W.W. Adams and R.M. Kelly, eds. American Chemical Society, Washington, DC. p. 1088-1121.

Michels, P.C., and Clark, D.S. 1997. Pressure-enhanced activity and stability of a hyperthermophilic protease from a deep-sea methanogen. Appl. Environ. Microbiol. 63: 3985-3991.

Michels, P.C., Hei, D., and Clark, D.S. 1996. Pressure effects on enzyme activity and stability at high temperatures. Adv. Protein Chem. 48: 341-376.

Miller, J.F., Shah, N.N., Nelson, C.M., Ludlow, J.M., and Clark, D.S. 1988. Pressure-temperature effects on the growth and methane production of the extreme thermophile *Methanococcus jannaschii*. Appl. Environ. Microbiol. 54: 3039-3046.

Miller, J.F., Nelson, C.M., Ludlow, J.M., Shah, N.N., and. Clark. D.S. 1989. High pressure-temperature bioreactor: assays of thermostable hydrogenase with fiber optics. Biotechnol. Bioeng. 34. 1015.

Mozhaev, V.V., Heremans, K., Frank, J., Masson, P., and Balny, C. 1996. High pressure effects on protein structure and function. Proteins. 24: 81-91.

Nelson, C.M., Schuppenhauer, M.R., and Clark, D.S. 1992. High-pressure, high-temperature bioreactor for comparing effects of hyperbaric and hydrostatic pressure on bacterial growth. Appl. Environ. Microbiol. 58: 1789-1793.

Ostendorp, R., Auerbach, G., and Jaenicke, R. 1996. Extremely thermostable

L(+)-lactate dehydrogenase from *Thermotoga maritima*: cloning, characterization, and crystallization of the recombinant enzyme in its tetrameric and octameric state. Protein. Sci. 5: 862-873.

Perler, F.B., Xu, M.Q., and Paulus, H. 1997. Protein splicing and autoproteolysis mechanisms. Curr. Opin. Chem. Biol. 1: 292-299.

Pfeil, W., Gesierich, U., Kleemann, G.R., and Sterner, R. 1997. Ferredoxin from the hyperthermophile *Thermotoga maritima* is stable beyond the boiling point of water. J. Mol. Biol. 272: 591-596.

Pledger, R.J., Crump, B.C., and Baross, J.A. 1994. A barophilic response by two hyperthermophilic, hydrothermal vent archaea-an upward shift in the optimal temperature and acceleration of growth rate at supraoptimal temperatures by elevated pressure. FEMS Microbiol. Ecol. 14: 233-241.

Privalov, P.L., and Gill, S. 1988. Stability of protein structure and hydrophobic interaction. Adv. Protein Chem. 39: 191-234.

Rice, D.W., Yip, K.S., Stillman, T.J., Britton, K.L., Fuentes, A., Connerton, I., Pasquo, A., Scandura, R., and Engel, P.C. 1996. Insights into the molecular basis of thermal stability from the structure determination of *Pyrococcus furiosus* glutamate dehydrogenase. FEMS Microbiol. Rev. 18: 105-117

Riera, J., Robb, F.T., Weiss, R., and Fontecave, M. 1997. Ribonucleotide reductase in the archaeon *Pyrococcus furiosus*: a critical enzyme in the evolution of DNA genomes? Proc. Natl. Acad. Sci. USA. 94: 475-478.

Robb, F.T., and Maeder, D.L. 1998. Novel evolutionary histories and adaptive features of proteins from hyperthermophiles. Curr. Opin. Biotechnol. 9: 288-291.

Robb, F.T., Place, A.R., Sowers, K.R., DasSarma, S., and Fleischman, E.M. (eds). 1995. Archaea: A Laboratory Manual. Cold Spring Harbor Laboratory Press, Cold Spring Harbor, NY

Russell, R.J., Ferguson, J.M., Hough, D.W., Danson, M.J., and Taylor, G.L. 1997. The crystal structure of citrate synthase from the hyperthermophilic archaeon *Pyrococcus furiosus* at 1.9 A resolution. Biochem. 36: 9983-9994.

Russell, R.J., and Taylor, G.L. 1995. Engineering thermostability: lessons from thermophilic proteins. Curr. Opin. Biotechnol: 6: 370-374.

Samarasighe, S.D., Campbell, D.M., Jonas, A., and Jonas, J. 1992. High resolution NMR study of the pressure-induced unfolding of lysozyme. Biochem. 31: 7773-7778

Schafer, T., Bonisch, H., Kardinahl, S., Schmidt, C., and Schafer, G. 1996. Three extremely thermostable proteins from Sulfolobus and a reappraisal of the 'traffic rules'. J. Biol. Chem. 377: 505-512.

Schellman, J.A. 1997. Temperature, stability, and the hydrophobic interaction. Biophys. J. 73: 2960-2964.

Starich, M.R., Sandman, K., Reeve, J.N., and Summers, M.F. 1996. NMR structure of HMfB from the hyperthermophile, *Methanothermus fervidus*, confirms that this archaeal protein is a histone. J. Mol. Biol. 255: 187-203.

Sterner, R., Kleemann, G.R., Szadkowski, H., Lustig, A., Hennig, M., and Kirschner, K. 1996. Phosphoribosyl anthranilate isomerase from *Thermotoga maritima* is an extremely stable and active homodimer. Protein Sci. 5: 2000-2008.

Stetter, K.O. 1996. Hyperthermophilic prokaryotes. FEMS Microbiol. Rev. 18: 149-158.

Summit, M., Scott, B., Nielson, K., and Baross, J. 1998. Pressure enhances thermal stability of DNA polymerase from three thermophilic organisms. Extremophiles. 2: 339-345.

Sun, M.M.C., Tolliday, N., Vetriani, C., Robb, F.T., and Clark, D.S. 1999. Pressure-induced thermostabilization of glutamate dehydrogenase from the hyperthermophile *Pyrococcus furiosus.*" Protein Sci. 8: 1-8.

Trent, J.D., Kagawa, H.K., Yaoi, T., Olle, E., and Zaluzec, N.J. 1997. Chaperonin filaments: the archaeal cytoskeleton? Proc. Natl. Acad. Sci. USA. 94: 5383-5388.

Trent, J.D. 1996. A review of acquired thermotolerance, heat-shock proteins, and molecular chaperones in archaea. FEMS Micro. Rev. 18: 249-258.

Vetriani, C., Maeder, D.L., Tolliday, N., Yip, S.P., Stillman, T.J., Britton, K.L., Rice, D.W., Klump, H.H., and Robb, F.T. 1998. Protein thermostability above 100°C: A key role for ionic interactions. Proc. Natl. Acad. Sci. USA. 95: 12300-12305.

Weber, G., and Drickamer, H.G. 1983. The effect of high pressure upon proteins and other biomolecules. Quat. Rev. Biophys. 16: 89-112.

Index

V

Books of Related Interest

Molecular Biology: Current Innovations and   1995
Future Trends
A.M. Griffin and H.G. Griffin (Eds.)

Internet for the Molecular Biologist   1996
S.R. Swindell, R.R. Miller, G.S.A. Myers (Eds.)

Gene Cloning and Analysis: Current Innovations   1997
Brian C. Schaefer (Ed.)

An Introduction to Molecular Biology   1997
Robert C. Tait

Genetic Engineering with PCR   1998
Robert M. Horton and Robert C. Tait (Eds.)

Prions: Molecular and Cellular Biology   1999
David A. Harris (Ed.)

Probiotics: A Critical Review   1999
Gerald W. Tannock (Ed.)

Peptide Nucleic Acids: Protocols and Applications   1999
Peter E. Nielsen and Michael Egholm (Eds.)

Intracellular Ribozyme Applications: Principles and Protocols 1999
John J. Rossi and Larry Couture (Eds.)

Current Issues in Molecular Biology   1999

Prokaryotic Nitrogen Fixation: A Model System   2000
Eric W. Triplett (Ed.)

For further information on these books contact:

Horizon Scientific Press   Tel: +44(0)1953-601106
P.O. Box 1, Wymondham   Fax: +44(0)1953-603068
Norfolk   Email: mail@horizonpress.com
NR18 0EH England   Internet: www.horizonpress.com

Our Web site has details of all our books including full chapter abstracts, book reviews, and ordering information:
www.horizonpress.com